THE 50 CLASS 50s

Their life and times from the London Midland through the Western Region to Network Southeast

By Bernard Mills

First Published 2024 - Copyright The Fifty Fund

ISBN 978-1-8384369-7-1

It had only been three months earlier the class had made their first appearance in Cornwall. It did not take long for them to become a regular sight in the Royal Duchy as they made a gradual inroad, sharing duties with the Class 52s they had been despatched to replace. 50 025 hurries through the cutting just to the west of Forder Viaduct near Saltash in charge of 1B65 1330 Paddington - Penzance. 22 June 1974

Contents

Contrast in liveries as 50 032+50 001 approach Plymouth Station passing the site of the former locomotive servicing point in the V of the former junction of the line to Millbay, 1345 Penzance - Plymouth 05 October 1981

Glossary

Abbreviations used in this book:

BR - British Railways/British Rail
DMU - Diesel Multiple Unit
EMU - Electric Multiple Unit
GWR - Great Western Railway
HST - High Speed Train
LMR - British Rail London Midland Region
LNWR - London & North Western Railway
NSE - Network Southeast
SR - British Rail Southern Region
TOPS - Total Operations Processing System
WCML - West Coast Main Line
WR - British Rail Western Region

The pictures in this book are all from Bernard's extensive archive. Many were taken before current film emulsions and processing techniques were available, so may not meet the standards we now take for granted in the digital age, but are still included here for historical interest.

Cover Photos

Front Cover Top: 50 017 runs along the Sea Wall approaching Parsons Tunnel. 15 July 1989
Front Cover Lower: 50 036 approaches Shotlock Hill Tunnel. 24 June 1974
Rear Cover: Treskelly Farm St Germans Bank, 50 010 leads 2C83, the 0911 Penzance - Plymouth on 7 July 1987

Foreword

It is now over thirty years since the Class 50 Fleet was withdrawn from service with British Rail. The Class always had a huge following especially in the 1980s and 1990s by which time they had been forgiven for usurping the hallowed Westerns or Thousands as we knew them. Even after withdrawal, that following has continued and there is possibly an even stronger one for the Class today as younger members, who were not privileged to witness the locos in service join those ranks of the elderly who did, now following them in preservation and even out on the mainline as they earn their keep on various stock moves or haul rail tours to diverse and distant parts of the network.

Bernard Mills is not only a professional railwayman, now retired, but also a renowned and well-respected Railway Photographer, particularly in the West Country. He has spent his whole railway career around East Cornwall and West Devon in what would have been the former Plymouth Division on the Western Region of British Railways. As such Bernard has two significant advantages over other photographers: -

- Firstly, he is well known by Rail Staff throughout the area and is therefore 'in the know' for any special or unusual movements likely to take place such as the weed killing train on the St Ives Branch or a Fifty on the Gunnislake Branch, as such he would often get prior notice from the local operations team of any notable movements about to take place, information he was able to share with fellow enthusiasts.
- Secondly, he is known to Signalmen, Permanent Way staff and Traincrew throughout the South West which enables him to gain access to the railway not available to the general public, and he can call on favours from drivers to wait for him to take the shot before departing, both of which are demonstrated on a few occasions in this book.

50 008+50 015 1350 Laira-Old Oak Common at Laira Junction returning D1015 to Old Oak Common depot after the Laira Open Day in 1991. The picture above says much about our ethos at Laira in those final years: *"Immaculate locomotives presented to their highest standard ready to delight their awaiting public."* 19 September 1991

The Class 50 story, depicted from the author's collection, is told in chronological order starting with their introduction in 1967 on the London Midland Region. By then electrification of the West Coast Mainline (WCML) had reached Crewe and there was a need for a powerful new Diesel Electric locomotive to take over from electric traction to haul the crack Anglo-Scottish expresses over the steep gradients of Shap and Beattock, hence the introduction of a new 100 mph locomotive capable of multiple working D400 to become Class 50 on renumbering. It is not appropriate to detail their history and the reasons for their design in this volume as they are well documented elsewhere.

From 1972 onwards as electrification of the WCML progressed, and the decision by British Rail to replace the Western Class diesel hydraulics (Class 52) with the Class 50s they were gradually transferred to the Western Region. Reference has been made of the frequent change of operating sphere of the Class but there are in fact two constants in their career on BR. The first 5 years was on the LMR and the next twenty plus on the WR. They had two principal homes. Old Oak Common in London and Laira in Plymouth, depots that were to remain their home throughout the rest of their career until the final days when the remaining fleet was all allocated to Laira.

Initially their main diagrams were Paddington to Bristol services, but as High Speed Trains (HSTs) were introduced on the Bristol and South Wales routes the Fifties took over the West of England Diagrams. So, Bernard located in Plymouth was well placed to capture the action of the Class throughout the West Country which also included North East South West services and also the Summer Saturday Holiday traffic to such places as Newquay.

In the early days of Class 50 operations, he was busy chasing his beloved Westerns and frequently a Fifty would appear instead of the Thousand, much to his disappointment. But following his motto *always take the picture and press the shutter, as its tomorrow's history*, thank goodness he did as we now have some early history of Fifties to savour. When the class transferred onto the Waterloo - Exeter route Bernard was still well placed because the locos would cycle through Laira Depot and so he was again in the right place at the right time.

From a personal perspective this book documents my own involvement with Fifties. I first became acquainted in 1969/1970 when staying with relatives in Penrith and can well remember double headed class 50's on Shap at Scout Green. Then in 1974 as an Engineering Trainee with British Rail I was seconded to Bath Road Depot in Bristol as the first Fifties arrived and cut my teeth as it were load banking and riding the class. Then in 1989 I was appointed as Area Fleet Manager Plymouth based at Laira Depot and was to oversee the final years and rundown of the class and eventual transfer of several locomotives into preservation.

As the rundown of the fleet increased the interest in the class grew and a number of rail tours were run initially using any available Class 50. In September 1990 the Taw Tourer railtour was run using 50031 and 50032 in very poor external condition. I felt that more effort should be made to demonstrate our pride and present the locos to the highest possible standard. So, as you will see picture on the last page, the Civil Engineer's Department authorised the repaint of two of their fleet. By May 1991 with funding from Rail Magazine 50050 was transformed into D400. The interest was immense and generated increased revenue as enthusiasts followed the retro loco on service trains. Such was the success that as the final few remained the decision was made to keep three locos for dedicated railtour use and so their life was extended for a further two years to run several spectacular tours seeing the class out in style. The end came along with the end of British Rail in March 1994 as the railway moved into the new privatised era.

I hope you get as much enjoyment from reading this book as I did in reviewing it.

Thank you Bernard for capturing a significant part of the class 50 story.

Geoff Hudson

Area Fleet Manager Laira Depot Plymouth 1989 - 1996

Introduction

The early days: just a couple of months old, D412 heads for London at Farington Curve Junction just to the south of Preston with 1A45 1141 Blackpool North - London Euston whilst Black 5 45330 has just come off the East Lancs line with loaded coal, 8P21 1050 Burnley Bank Top Sidings - Fleetwood Wyre Dock working. The bridge in the background is Bee Lane and just beyond is Farington Curve Junction Signal Box. In the foreground is the colour light home signal for Skew Bridge Junction Signal Box. The area was taken over by the Preston Power Box in November 1972, and doubtless this is a much changed scene today. 06 June 1968

British Rail under its Modernisation Plan of 1955 dived headlong in replacement of its steam fleet in to a number of diesel locomotive classes, many of which ordered "off the shelf" proved to be unsuccessful. As the 1960s wore on, it was realised a number of mistakes had been made. As the last steam locomotives went to that big engine shed in the sky, around the same time so did some of the diesels meant to have replaced them, and the death warrant for the diesel hydraulics of the Western Region had been signed. However, the need for a more powerful - and reliable - type 4 locomotive class had been identified and some consideration was given to the matter. The result was what would become the class 50 fleet, which can ascribed as the final flowering of the 1960's headlong rush into dieselisation, the first D400 appearing as 1967 drew to a close. Certainly they were not greeted with open arms by the enthusiasts of the day. In that last year of BR steam where it was concentrated in the North West, this was the territory to which they came, unwelcome as steam took its final gasps. When in 1972 the first moved south to the Western Region, here too they were not exactly welcome as their coming heralded the demise of the much loved Class 52 Westerns, a process that would take the better part of five years. Yet they grew on us and as their front line work gradually eroded by the spread of the High Speed Train and then the Sprinter, their popularity and following had grown to the point that they were positively welcomed on taking charge of the Waterloo - Exeter route in May 1980.

Their final fling probably remembered best for their exploits serving Network Southeast as devotees found much solace in their workings especially those from Paddington to the Thames Valley and

To the Western Region: By the long hot summer of 1976, the Class had become firmly established on Paddington - West of England workings, sharing and gradually taking over more of the duties from the beloved Westerns. Passing the iconic Horse Cove at Dawlish, a rather scruffy 50 032 heads west with 1B65 1330 Paddington - Penzance. 25 August 1976

Waterloo to the very outpost of NSE in rural Devon. Their work was further eroded as HSTs appeared on some Cross County services and the Sprinter revolution of the late 1980s as local services in the West Country for example saw the end of locomotive hauled local trains. As the class became expensive to maintain their use diminished, they finally surrendered the Waterloo trains in May 1992 to the Class 47s. It would be in March 1994 with just over a quarter of century of BR service the class bowed out. Like the Westerns and the Deltics they had gathered a huge following which persists to this day. A remarkable 18 members of the class survived the trip to the scrap yard and enter preservation, at present in varying states of order and condition. In the era of the privatised railway, they returned to the main line on rail tour duties and in recent years a few have even returned to capital stock for various duties such as stock moves and hauling the Rail Vac around.

I have often thought their career was very much a series of stop gap moves. Introduced in 1968 to bolster motive power requirements on the WCML north of Crewe, their heyday here would be over six years later when the Crewe (Weaver Junction)-Glasgow electrification was completed. The Western Region found them ideal for the Paddington-Bristol expresses, but here again their heyday was short as HSTs took over most of the diagrams from October 1976, although there would be a short resurgence on the route in 1980 due to HST problems. Superiority on the Paddington-West of England route would be shaken as from 1979 HSTs came on the scene. Tenure on the Waterloo-Exeter and associated workings would be a dozen years.

Once class 50 dominance was established on the WR, sadly their performance did not match the required standards. From 1978 to 1980 a complete refurbishment of the Class was undertaken at Doncaster Works. Even with the advent of the High Speed Train the WR needed a reliable class of

The brief heyday of the Class on the Paddington-West of England trains: Stormy skies over the Royal Albert Bridge at Saltash as 50 033 heads away from Cornwall into England with 1A77 0838 Penzance - Paddington. 25 November 1978

locomotives, the Class 50 would remain the dominant motive power for the Western Region for the 1980s locomotive hauled trains - of which there were still a substantial number - working the Paddington - Thames Valley, Worcester, Cheltenham and Birmingham routes as well as Cross Country services penetrating the LMR at Birmingham from the west. They would fill the gaps where there was an insufficient number of HSTs to cover all requirements like for example the odd Paddington-Bristol turn, Jumbo trains to bolster the West of England and until the close of the 1987 season they went on Summer Saturdays where HSTs could not. Their exploits on the heavy expresses on the steeply graded and tortuous curves of the Par - Newquay branch were legendry, and many were the enthusiasts who enjoyed this bonanza. Their final transformation would again be short as Network Southeast was created on 10 June 1986 to provide a unified trademark for all suburban and local working under one banner basically, with what is best described as some journalistic licence like all the way to the East Devon fields on the edge of Exeter, the London commuter area. This brought to some of the class a new and colourful livery, but their duties under NSE would be just for six years. Employment had been found on parcels and freight duties, but with the Sprinter revolution of the late 1980s, a backward move if there ever was one, coupled with further erosion of work and they were getting expensive to maintain, the first withdrawal would come in February 1987 with that of 50 011. It would take a further seven years for the class to pass into BR history as their farewell rail tour ran on 26 March 1994.

So with Class 50s still able to be found on a number of heritage railways and their occasional exploits on to the main line, the class survive as their fan base continues to grow. This book does not set out to be a detailed history of the Class, to discuss technical details or the various liveries they carried. Mine is a photographic record of seeing them over their 26 year life span in front line BR service from one end of England to the other. Like all the other photographers in the picture, this

The Class 50's became very popular rail tour engines well before their demise. Were you here in a scene which would throw the modern day Health and Safety brigade and the Office of Rail Regulation into convulsions? In contrasting liveries 50 025 and 50 040 pose for the assembled gallery, most of which standing in places now totally taboo, during a photographic stop on 1Z44 0550 Plymouth - Doncaster at Mexborough for a visit to Doncaster Works. The bonus of a double header was in one direction only, 50 040 had recently been released from Doncaster Works after refurbishment but had to return for rectification of some minor problems, the operating authorities saw the rail tour as the perfect way to get it back there. 26 July 1981

would be taken in the pre digital era; in my case on 35 mm colour slides of varying film over the years ranging from Ektachrome until the early 1990's when the improved German version of Agfa came on stream. Until the summer of 1987 I used a Praktica Nova with a Tessar lens before acquiring an Olympus OM1. Thus these are pictures before the quality of what we take for granted in today's digital era. All are scanned slides enhanced by the blessing of Photo Shop, if only we had that back in the days of film, and the latter did tend to be temperamental at times with colour variation etc, and some have stood the test of time better than others. I hope the reader will find the pictures acceptable. Other than the two pictures taken at Greenholme in August 1968, part of my collection and selected because they project better lighting conditions than those encountered on my visits to this part of the world at the time, they are all my own work. I must thank Barry Jones and Roger Geach for their help with scanning and picture processing, Jeremy Clark for guiding me through the mysteries of the software, Ron Herbert who worked in Preston Control in the era of the Class 50's in those parts for assistance and identifying individual train workings in the North West. Phil Prosser for sharing his knowledge of the Warrington area, Dave Mitchell for help with the early days of the Class on the Western Region, Colin Marsden for further information and last but not least that master of Class 50 knowledge and experience Geoff Hudson. I must also thank Joe Burr and the Fifty Fund for their faith in trusting me and producing this high quality publication which I hope all will enjoy. Any errors will be down to me, but with a rigorous editing and fact finding process I hope these will be very few and far between.

Chapter 1: The Early Years

My first sighting and photograph of a Class 50: An interloper to one of these last Cathedrals of Steam, brand new D414 at rest on Lostock Hall Shed in company with Black 5 44942 and friends just three months before they would breathe their last. D414 had been despatched to Lostock Hall for crew training, men of steam being taught new skills, for many a new way of life, for others a redundancy notice. The practice of stabling shiny new diesels on steam sheds with all the soot, dust and grime (such aromas a joy to savour) liable to spoil the paint work, and filter in to the mechanics, perhaps not ideal for the new order. 18 May 1968

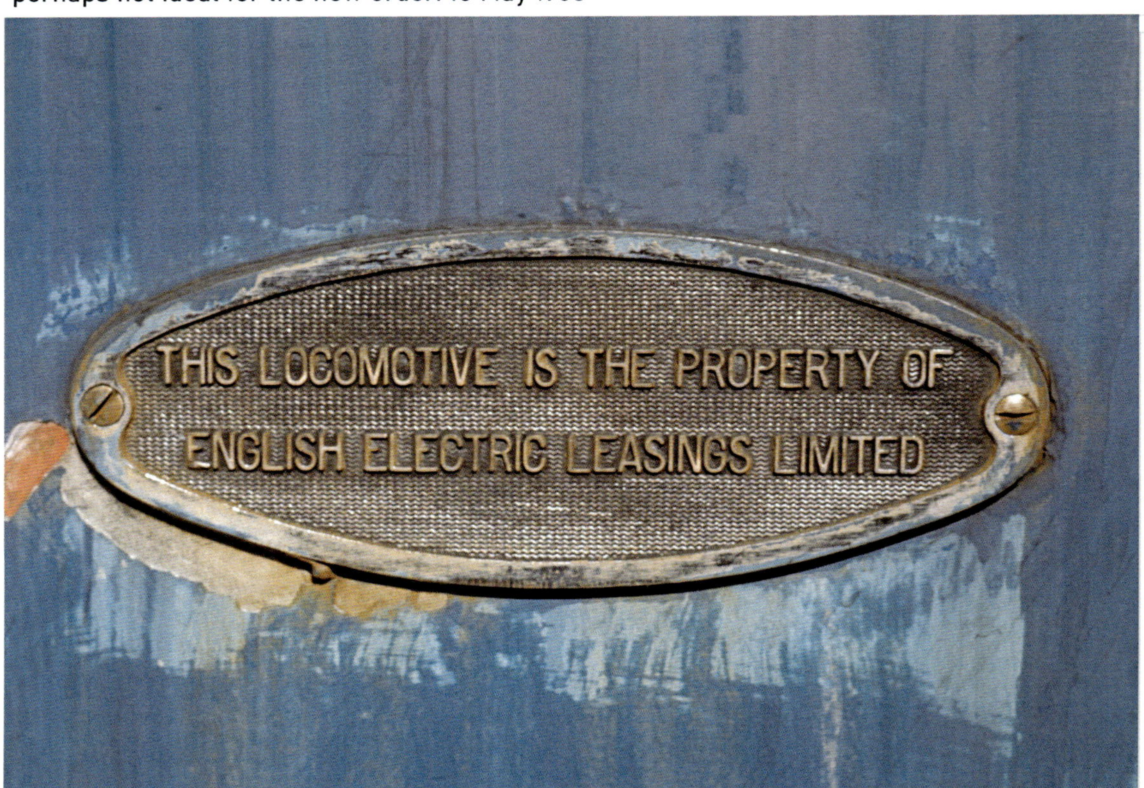

The first of the class D400 was introduced in October 1967, and just over a year later the last D449 was delivered in November 1968. They were built by English Electric at the Vulcan Foundry, Newton-Le-Willows. In a departure from the normal practice of the time, the fleet remained owned by the builders, and leased by BR for their first decade in service. The plaque carried by each member of the class indicating ownership as seen attached to 50 001 at Newquay. 03 September 1977

About three hours on from the previous picture, my second encounter with the class this time at Carnforth Shed as a brand new D419 stands behind D8506 in a veritable line up of the diesel motive power of the day, steam and diesel sharing the same facilities as to the right Black 5 44877 takes centre stage. This outing was a Plymouth Railway Circle organised weekend tour of the remaining steam sheds in the North West. I must admit my visits to the area at that time were to put as many steam engines as possible in the lens, any depicting the old and the new forms of motive power considered as a future historical bonus. These new fangled diesels would be here forever would they not?
18 May 1968

In that strange week between the last workings of the last BR steam on normal services and the famous 15 Guinea Special on 11 August 1968 which would mark the end of steam on BR, D428 climbs away from Tebay at Greenholme 1S62 1140 Birmingham-Edinburgh 07 August 1968

As the steam age passed, the brand new class 50s had become established as the prime motive power north of Crewe on the WCML as D405 followed D428 above working 1S63 1205 London Euston-Glasgow Central. Only twelve months previous I had stood in these parts enjoying the green fields of Shap photographing and enjoying steam conquering the Northern fells. There were not any class 50s to be seen that summer. The winds of change had blown. 07 August 1968

My first visit to the North West after the end of steam in August 1968 was actually on the way to sample the summer Saturday delights of the North Wales Coast Main Line, travelling up from Plymouth over night on the 0115 to London Paddington behind D1059. After some early morning photography of the comings and goings at Paddington, it was over to Euston for 1P18 0905 to Windermere, which I rode as far as Warrington Bank Quay changing for 2J61 1210 Chester-Manchester Victoria service to Newton-Le -Willows. Return to Warrington was by foot by way of various photographic locations including Winwick Junction. After a meal and some beers, the 2005 ex Manchester Victoria 4 car DMU brought me into Chester at 2119 for a very welcome bed. The stamina of youth, much younger then and of course there was much more of interest to see on the rails. As for direct trains to London to Windermere, these are now a memory .Above we see D449 backing on the 0905 from Euston, taking over from E3113, a typical scene of the era of the change of motive power at Crewe. 10 July 1970

From May 1970 to allow accelerated timings between London and Glasgow, on certain trains double headed working in multiple Class 50's were rostered north of Crewe, This would last for only four years until the route was fully electrified, indeed for the last twelve months the motive power change would be at Preston.

A view that is still possible but without the signals and not entirely spoilt by the electric wires installed here in 1972, D430+D415 are seen on the outskirts of Newton Le Willows, Southwell Road with 1M20 1000 Glasgow Central-Euston Royal Scot. The train diverted this way off the normal route for engineering work. The M6 Motorway runs through here now.

After a fair walk and later the same day, the classic shot from Alder Lane Bridge as just after passing the place of their birth, D443+D414 pass Winwick Junction with 1M35 1400 Glasgow Central-London Euston. Due to engineering work, the train had been routed from the normal WCML route leaving the latter to take the original route at Lowton Junction, then joining the Liverpool & Manchester railway near the Huskisson Memorial heading west through Newton Le Willows Station. A mile later it squeezed around the tight curves that circumnavigated the gas holder seen in the background behind the train and headed south past the Vulcan Foundry to rejoin the WCML. The line to the right of the train is the "new" direct Lowton Junction-Winwick Junction route opened 1 August 1864 saving the need for trains to pass through the congested Newton-Le-Willows station. This is still the normal route today for WCML expresses. To the far right behind the signal box an unidentified class 40 can be seen on an engineering train. It is a much changed scene here today with the electric wires, and the Vulcan Foundry which made a superb back drop behind the train closed in 2002 and has subsequently been demolished. No trace of it remains today, the site now a housing estate. A place etched in to the history of the class 50s. 10 July 1970

Looking south from the Alder Lane Bridge, the going away shot under threatening skies as D443+D414 speed on towards Warrington with 1M35 1400 Glasgow-Central-London Euston. Note the rather fine signal gantry for northbound trains, this the Winwick Junction down home signal, and the over bridge beyond our next port of call.
10 July 1970

A short walk to view the scene at Winwick Quay, sadly the sun did not oblige for the fine sight of D438+D436 hurrying past with 1S83 1605 London Euston-Glasgow Central. Note the unusual combination for the Winwick Quay Signal Box Up distant signals,, the Up Goods line a semaphore distant standing beside the colour light version for the Up main line.
10 July 1970

Electrification of the West Coast Main Line invariably involved weekend diversions. Diverted from Crewe via Chester, on 11 July 1971 D405+D412 with 1F19 1010 Euston-Liverpool disturb the peace and quiet of the rural junction at Helsby as they head for Acton Grange Junction and Warrington along the route formerly a part of the Birkenhead Railway taken over by the GWR and the LNWR on 1 January 1860. To the right are the platforms for the also still extant branch to Ellesmere Port and Hooton, also part of the Birkenhead Joint. The LNWR signal box can be picked out just to the right of the first coach. 11 July 1970

The going away shot of 1F19 taken as much for the remarkable and unusual parallel co-acting arms signal for the junction in the down direction as well as the rare motive power passing through. Although sited on the other side of the track due to the clutter on the down side, the lower arm is in the main signal; the upper arm is a repeater which would give the footplate crew sight of the signal from beyond the over bridge in the background. Remarkably Helsby remains an oasis of mechanical signalling; sadly this almost unique specimen has been replaced on the same spot by two parallel standard LMR arms but only at the lower normal lower level. No doubt the removal of the telegraph poles and wires now provides a clearer view for down train crews here of the state of the signal. 11 July 1970

WCML trains diverted between Crewe and Acton Grange Junction took the North Wales line to Chester where a reversal to join the LNWR/GWR Joint line to Warrington was necessary. To avoid running the engine(s) around in the platforms, the procedure was to turn the whole train on the triangle at the west end of the station, and this the operating department got down to a fine art. The inbound train would run through the station and take the curve towards Birkenhead, reversing at Chester No 5 signal box to take the curve to Chester No 6, reversing again taking the confused passengers through the station for the second time in 10 minutes. In a picture full of interest, D426 seen at Chester No 5 as if heading for Birkenhead with 1P31 1430 London Euston-Carlisle with behind the circa 1957 built diesel shed hosting a veritable selection of Type 2s awaiting their Monday morning duties. The adjacent blue ended building is the former GWR shed, in use after its closure to steam as a DMU depot. Moments after this picture was taken, D426 would propel its train around to Chester No 6 by the curve seen in the immediate foreground. 11 July 1971

Having propelled their train from Chester No 5 to Chester No 6, D418+D419 working 1F19 1010 London Euston-Liverpool Lime Street have now reversed and take the curve from the North Wales main line towards Chester General Station. On show is a magnificent collection of semaphore signalling with the back ground dominated by the former GWR steam shed with miscellaneous DMU's parked up alongside. 19 September 1971

The purpose of our outing that day was to find the far end of the diversionary route where the trains would leave/join the WCML, Acton Grange Junction just to the south of Warrington. This a rare chance to record main line trains on the limit of the LNWR/GWR joint line at this location. Heading south is D438 with the lengthy 4A56 1340 Heysham Harbour-York Way FLT Freight Liner service. The class 50s were not a common sight on such trains. 19 September 1971

Acton Grange Junction, the most northerly point of the GWR through its joint ownership with the LNWR of the former Birkenhead Railway. The view from Moore Lane as D446 heads south towards Chester with the diverted 1A63 1531 Blackpool North-London Euston. To the left, D5254 stands by the signal box on the WCML with an engineering train, the masts for the electric wires in position. The tunnel is the former route of the LNWR/GWR joint line from Chester, abandoned after major works and re alignment of the railways here (which caused much consternation to the LNWR) for the construction of the Manchester Ship Canal opened in 1895. The scheme included the substantial bridge necessary to take the railway over the latter is behind the train. 19 September 1971

A further view just down the road from the location on the previous page on the LNWR/GWR Joint Line deviation of 1895 as D415 heads south towards Chester with the diverted 1605 Blackpool North-London Euston. Acton Grange Junction with the Manchester Ship Canal Bridge is in the background, the partially wired WCLM is to the left. Trains bound for Chester take a right hand turn just beyond this point to rejoin the original route and pass under the WCML. 19 September 1971

An overnight stop at Carlisle on the way to the Penrith-Keswick branch the following day provided an opportunity to catch up with a train that had fame both good and bad. The West Coast Postal was probably the most important train that plied the route. The down Scottish bound working is forever immortalised by the film Night Mail of 1936 by the poem of W H Auden, in my opinion the most romantic piece of railway writing ever written. The Up working is more infamous being the subject of the Great Train Robbery of the night 7/8 August 1963 at Bridego Railway Bridge, Ledburn in Buckinghamshire. In reality the working was 1M44 1905 Glasgow Central-London Euston, also known as "The Up Special TPO" (Travelling Post Office). D421 observes the 15 busy minute stop at Carlisle, to the right D5352 stands awaiting its next turn of duty. The note book records the exposure on 64 ASA Ektachrome film as 90 seconds at f11. 03 December 1971

The only credible picture on the main line the following day due to poor weather was this of D404 arriving at Penrith with 1P16 0925 Crewe-Carlisle. An image that brings back the aura of the railway of old. Since the end of steam, the only thing had had changed was the diesel on the front of the train. To the left is the coal yard still being used by the merchant, but sadly the coal came by road as the yard closed to rail traffic in January 1971.
Notice to the left of D404 a fine lattice work post co-acting arm signal on the up main line, and to the right of the engine the rear of a banner repeating signal. This is a repeater for a signal whose view is obscured by the curved platform canopy of Penrith station. Look to the right of the penultimate coach of the train for the signal that "never was" covered by an X. This was erected for the new WCML Carlisle Power Signal Box project to control trains departing from the up and down passenger line for the Keswick Branch and would have been Signal CE 205. But the branch closed on 6 March 1972, and Carlisle took over on the week-end of 3-4 March 1973. 04 December 1971

A brief visit to the Border City accompanying my late parents this far on their way to a holiday in Bonnie Scotland, making sure they were on the right road!! D434+D437 both looking smart make a colourful sight observing the Carlisle stop with 1S63 1205 London Euston-Glasgow Central, 04 July 1973

The amount of Parcels and GPO Mail traffic conveyed by rail was still very substantial at this time, providing any number of mail and parcels workings the length and breadth of the land. With a motley collection of vans in tow – typical of what could be seen in this era – D442 waits with 3A65 1700 Carlisle-London Euston. Note immediately behind the engine the two fine ex LMS specimens. 04 July 1973

My next encounter with the class 50s would be on home soil, my first picture of one in Plymouth, the sign of things to come. Laira Shed on Christmas Day with D1029 Western Legionnaire at rest, to the left D400 recently arrived for crew training purposes. 25 December 1973

As the electric wires reached Glasgow, the initial employment of the class 50s was reaching its conclusion. BR policy was to keep the class intact to one region. With the need for a strong class of engine to finally replace the class 52 Westerns and bolster its resources, the move south to the Western Region was the logical move.

To the Western Region: I have through the good offices of Dave Mitchell utilising gen from the Railway Observer, Modern Railways and the note books of Chris Parsons, Mike Hunt and Sean Greenslade, further information from Roger Geach and my own observations, I compiled some notes of the transfer from the class from the LMR, and these are recorded here out of interest:

- Week ending 21 October 1972 D400 was allocated to 82A Bristol Bath Road depot.
- 1 December 1972 D400 worked an 8 coach crew training run from Paddington.
- 15 January 1973 D400 worked the 1712 Paddington-Weston-Super-Mare; this is thought to be the first passenger run on a weekday from London Paddington.
- 7 June 1973 D400 worked 4B10 1558 Bristol Temple Meads-Plymouth Parcels; this is believed to the first trip over the South Devon Banks for the class.
- 23 July 1973: D401 the second to be transferred to the WR arrived in Bristol being towed dead by Peak D142 (No trained crew available), It worked the 1515 to Paddington the following day.
- On 28 July D400 became the first member of the class to be seen at Weymouth on an excursion from Bristol TM. Sadly it disgraced itself by failing on the return journey between Dorchester South and Maiden Newton, and is believed to have been pushed by the following train to Yeovil Pen Mill arriving there 1½ hours late. Here D1600 was waiting having been worked down light engine from Westbury to take the train forward. Not a happy homecoming for the day trippers.
- November 1973: Reported in the Railway Observer if available D400/401 share this duty: Mondays to Fridays 0805 Bristol TM-Paddington and 1145 return. 1515 Bristol TM-Paddington and 1845 return. Saturdays: 0807 Weston Super Mare-Paddington and 1415 return, Sundays 1745 Bristol TM-Paddington and 2245 return. D401 observed on the 0805 on 12 November 1973

Close up view of D400 on Laira Shed, parked up in the sand road where it was out of the way, as at the time it was used for artisan training rather than driving training for much of the time. The contraption on the roof of the Sand House to the left of the engine is part of the extractor system for the sand drier 25 December 1973

- 11 November 1973:D400 passed through Newton Abbot bound for Laira Shed, Plymouth, for crew training. Dainton Bank proved too much for it as within thirty minutes it had returned! Having spent four years doing battle with the northern fells, one would have thought it would be well used to hills. After stopping at Newton Abbot West Signal Box it proceeded through the station and was taken to the stabling point by D4185. First class 50 to be reported on shed at Newton Abbot.
- 22 November 1973:D400 has been crew training from Laira Depot. On this day the 0800 Bristol TM-Penzance arrived at Exeter St David's 40 minutes late with class 25 D5230 towing dead Peak 46 008. The 25 then placed the 46 in the centre road and coupled back on to the train. Meanwhile D400 was waiting near the stabling point to return to Plymouth on a crew training run, but instead was put on the Penzance bound train. Although it was only a seven coach formation, both locomotives worked the train to Plymouth. This is thought to be the first passenger train working in Devon for the class 50s.
- In November 1973 D402 replaced D400 as the crew training engine at Bristol Bath Road Depot and was observed at Exeter St David's on 1 December 1973 after working a down parcels train. Meanwhile D401 spent the latter half of December 1973 and all of January 1974 on loan to Old Oak Common (London). Crew training continued apace as more members of the class migrated south from the LMR, this coinciding with the renumbering of the class under the TOPS (Total Operations Processing System) introduced by BR in 1972 and still in use today. D400 was renumbered 50 050 at Laira Depot on 27 February 1974.
- A further milestone occurred on 5 March 1974 when 50 027 on a light engine test run from Plymouth to Penzance became the first of the class to be seen in Cornwall, and this then became the Penzance crew training engine. Their association with the Royal Duchy would last for two decades to the month. As the electrification to Glasgow was completed, more of the class were despatched south. During the week commencing 28 April 1973 the balance of 27 class 50s for the WR were despatched from Crewe to Bath Road and Laira bringing the number of the class on the WR to 35. They travelled south in pairs or trios via Craven Arms and were immediately put to work on Paddington/Bristol and West of England duties. As the new timetable came into force from 6 May, about two thirds of Paddington-Torbay/Penzance services were booked for either class 47 or class 50 haulage. This was the first day of class 50 diagrammed workings in the West Country. Much to the delight of many, a number of services on the West of England, South Wales and Birmingham routes from Paddington were still allocated to the class 52 Westerns. The winds of change were blowing.

By the summer of 1974 the class 50s had become a familiar sight on the WR and spread their wings to the West Country gradually infiltrating the work of the class 52 Western Hydraulics, as crew training and transfers south became effective.

Being fitted with ETH (Electric Train Heating) they were able to allow the gradual introduction of Mark II air conditioned stock. The non such fitted stock was still very much the order of the day as 50 044 is seen approaching Dawlish Warren station working 2B31 1655 Paignton - Exeter St David's. 14 June 1974

The LMR retained 15 class 50s for residual duties for a couple of years, their prestige trains rushing passengers to and from Scotland over the WCML now in the hands of the class 87s. The sight and sound of the class 50 still could be heard (in much reduced numbers of course) over the Northern fells, not necessarily on the metals of the WCML but also those of the former Midland Railway route namely the Settle & Carlisle Railway. On a glorious sun lit morning with hardly a cloud in the sky, and believe me one does not encounter many of these in this part of the world, 50 035 heads south at Ais Gill Summit with 8G80 0805 Carlisle-Bescot goods. This was still the era of the traditional goods train with its variety of short wheel base wagons and vans, and running such trains over the Settle and Carlisle kept the line over Shap free for the brand new electrics and their very fast trains. 24 June 1974

Sequential numbering: Half an hour after 50 035 had been captured at Ais Gill, a move the short distance south to the top of Shotlock Hill Tunnel, the highest in England, saw 50 036 heading south with 7V57 0835 Carlisle-Severn Tunnel Junction.

Again a fairly lengthy goods train with a varied load was nothing out of the ordinary at this time. As with the picture at Ais Gill, the back ground is dominated by the majesty of Wild Boar Fell.
24 June 1974

Back home from my wanderings to the Northern hills now I could see class 50s going over the South Devon hills. I must admit at this time my primary aim was to place as many class 52 Western's in the frame as possible. In retrospect it was an interesting time in the West Country with a variety of motive power to be seen and enjoyed. Having conquered Dainton Bank, 50 001 is about to enter Dainton Tunnel with 1B65 1330 London Paddington-Penzance, a train now regularly producing a class 50. Note opposite the rear of the engine on the up line a relic of the steam age, the stop board reminding train crews of unfitted freight trains to stop and pin down brakes for the descent towards Aller Junction. Then neatly in order the signals are the Dainton Tunnel Signal Box up starter with the Stoneycombe Up distant opposite the rear of the train. 03 August 1974.

High tide at Hayle: Class 50 in the far West. On a lazy summer Saturday afternoon 50 007 heads over Hayle Viaduct with 1V10 0901 Derby-Penzance. Mark 1 stock and summer Saturday extra's. Remember them? The train was in effect a relief to the down Cornishman. 17 August 1974

50 027 hurries through Tiverton Junction with 1B53 1226 London Paddington-Paignton, taken from the bridge by Lloyd Maunders meat processing plant where photographers were not always welcome. By this time the station was a junction no more, the branch to Tiverton served by the platform to the left behind the signal box had closed to passengers in September 1964, that to Hemyock served by the south side of the down platform a year earlier. At this time the line to Hemyock was still open for milk traffic, this ceased in 31 October 1975. The station would close on 12 May 1986 when replaced by Tiverton Parkway, two miles up the line on the site of the former Sampford Peverell station, the buildings demolished in April 1991. The line had been quadrupled through here in the 1930's, the loops survive in use known as Tiverton Loops controlled from the Exeter Panel Signal Box. 31 August 1974.

50 003 pauses at Bodmin Road with 1V76 0920 Liverpool Lime Street-Penzance, the station became a junction on 27 May 1887 when the GWR opened its branch line to the town itself. Further significant changes came to the station in the Class 50 era, when it was renamed Bodmin Parkway on 4 November 1983 as part of the newly branded Cornish Railways when some degree of autonomy was granted to local management. On the 20th of the same month a farewell special traversed the Bodmin branch as it closed completely. The branch to Bodmin General was reopened by the Bodmin & Wenford Railway on 17 June 1990. In the same year, 50 042 Triumph was withdrawn at Laira Depot, it made its way to Bodmin General, and after some restoration work came into service in 1993. This popular locomotive has made the Bodmin & Wenford a place of pilgrimage for Class 50 devotees. 03 January 1975.

By 1975, the class 50s had become an integral part of the WR scene. A useful train for afternoon photography in that era was 7B33 1400 Exeter Riverside-Truro Goods, seen here with 50 003 at Bolitho Viaduct immediately to the east of Liskeard. The consist on this occasion: loaded china clay for export via Fowey Docks, more than likely from Heathfield and North Devon. 20 March 1975

March 1975 is a good time to conclude my look at the early years of the class 50s: from their introduction in the North West and exploits on the LMR to their migration south to the WR. On a personal basis, my photography in our fair land would cease for almost until the end of 1975 due to my six month secondment to Brit Rail Travel International in New York. After that came a tour to Asia exploring the railways of northern India and Thailand.

I sign off my look at the early days of the class 50 with a location difficult to reach. A clean and smart looking 50 025 coming off Liskeard Viaduct, the latter not immediately standing out in this difficult to reach view point; spot the guarding parapet rails opposite the coaches, 1A79 1440 Penzance-London Paddington. The over bridge in the back ground will be familiar to many as this spans the platforms of Liskeard station. So after my globetrotting, it will be time to head to Chapter two and have a look at each of the 50 class 50s. 20 March 1975.

Chapter 2: The 50 Class 50s

Profile side end view of 50 050 at Falmouth waiting to depart with 1A79 0910 to London Paddington, 08 May 1976, the first summer Saturday of the 1976 season, and the first visit of a Class 50 to this Cornish branch line terminus.

As 1976 dawned I had returned back to Plymouth from my travels to resume normal photography. The year is best remembered on two counts, one the long hot summer and drought, the other this was the last full year of the Class 52 Westerns in traffic. By the end of spring that year, the transfer of all the Class 50's to the WR had been accomplished. Back tracking in time for a moment, I had previously mentioned the renumbering scheme under the introduction of the TOPS system by BR, which was applied to all members of the class over the first four months of 1974. In simple terms, the TOPS system required all locomotives to be renumbered with the first two numbers being the actual Class number. i.e. 50 followed by a space and the commencing with 001 all the locomotives in numerical order. Simple enough for the 50s, D401 became 50 001 and so on. For some reason the system could not cope with the first engine in the sequence starting with the last three numerals being 000, so D400 was relegated to the bottom of the list becoming 50 050.

After the demise of the Class 52 hydraulics at the end of February 1977, there is no doubt the Class 50 became the flagship of the Western region. The following year 1978 is a notable one for the Class on a number of counts. As previously mentioned, when built, they were the subject of a 10 year lease from English Electric Leasings Ltd to the British Railways Board, the latter not wanting to invest any more money in new locomotives at the time. It is a bit of a grey area from when the leasing period ran. Papers at the Public Records Office in Kew show that they started on the day of delivery and when the certificate of insurance was issued. This would mean for D400 from 09 October 1967 through to the final member of the class delivered D449 on 12 December 1968, which have seen them all become BR property between October 1977 and December 1978.

However, the BRB annual accounts show all 50 were BR property in April 1978, but no paper work can be found which actually records the transfer. For the accountants it was likely 04 April 1978 at the end of the EE (English Electric) accounting year. Also no price has ever been available as to what

A time of transition as the now preserved D1015 Western Champion and 50 011 engage in some shunting moves at Laira Depot. At the start of 1976 BR dispensed with the requirement for head codes to be displayed. A subsequent unofficial move adopted by many depots was to display the locomotive number, or part of it, in the now redundant head code panel box. It fits well for 1015, for the Class50's one of the 0's was omitted to identify the engine, the abbreviated version shown to good effect on 50 011. This was not Tintagel Castle resurrected!! 07 November 1976

price BR paid for them. The suggestion by Mike Scott at EE at the time was £48 million for the full fleet of 50 locomotives plus spare engine units and bogies. This does give us an insight in to the sort of money, huge sums even at that time, required to buy motive power. My thanks are due to Colin Marsden for this information and permission to publish.

Now in BR ownership and very much the flagship of the Western Region, the idea came about to name the Class. It was decided with the long association between the railway and the Royal Navy that resurrecting the Warship brand for want of a better word would be appropriate, with a different theme to that applied to the former D600 and D800 Diesel Hydraulics. Commencing with 50 035 Ark Royal on 17 January 1978, with a couple of variations 50 049 Defiance being the Torpedo Training School at Devonport Dockyard and 50 050 Fearless whose namesake played a major part in the Falklands Task Force of 1982 liberating the Islands from Argentinean occupation, the fleet were named after Royal Navy Warships with distinguished records from both the First and Second World Wars. It was a wise move, each engine gained its own special identity and just somehow the "Warship" tag was most befitting. It would take until 25 September 1979 for the naming of the class to be complete.

Having spent a considerable amount of money firstly hiring the Class for a decade and buying them outright, the new owners were concerned by the Class being dogged by a number of problems ranging from minor faults to serious failures resulting in poor availability and performance. It was decided in 1978 the whole class would be despatched to Doncaster Works over a period lasting from early 1979 to the latter part of 1983 for refurbishing. Without going in to too much technical details which is really beyond the scope of this book, this involved removing things like slow speed control

The refurbishment of the whole class 50 fleet was undertaken at Doncaster Works from 1979 to 1983, and was the most comprehensive undertaken by British Railways to any locomotive type. The already mentioned, a 50 hauled rail tour from Plymouth to Doncaster included a visit to "The Plant" allowing for those of us living in the West Country easy access and take photographs, which of course included the Class 50 refurbishment programme.

In a corner of the Crimpsall Shop, undergoing the "treatment", 50 031 is in stages 2/3 of the refurbishment and looks somewhat forlorn; missing buffers, multiworking cables and cab windows, and with part patched bodywork and off its bogies - so resting on the work's accommodation stands. Work on some of the more obvious refurbishment changes has already been done, such as the hole for the new headlight in the center of the front below the cab windows. Alongside, and somewhat dwarfed' is diesel mechanical shunter 03 158.

Under the 1994 railway privatisation the locomotive side of Doncaster Works became the property of Bombardier. Sadly as the railway market and requirements for heavy overhauls changed, and this workshop was closed in 2007. The impressive and historic Crimpsall Workshop was demolished in 2008, and just as where the class were born at the Vulcan Foundry, the Crimpsall site is now covered in housing. The carriage and wagon side of the operation at Doncaster Works survives under Wabtec. A fall from grace one may say. 26 July 1981

and rheostatic braking which was rarely used and replacing the air take fan arrangement and fitting of high density head lights which did rather change their appearance, and this will be apparent in the survey of the whole class in the following pages. Also apparent will be the changing liveries - another wide ranging subject whose detail is beyond the scope of this book. In general the refurbishment was initially money well spent. The last major event in the life of the Class was the creation of Network Southeast in 1986. This would see some of them repainted in a very distinctive and colourful livery, briefly red, white and blue sums it for now. As the decade wore on age was catching up with the Class, they were getting expensive to maintain and the first 50 011 was withdrawn from service 24 February 1987.

There now follows a pictorial survey of the whole class. Unlike the pictures in Chapter 1, and as they will be in Chapter 3, they do not follow chronological order; instead these follow in numerical order covering their time after the whole class had been transferred to the Western Region in 1976, taking us far and wide around the WR and a few beyond as well, showing how they earned their keep doing the bread and butter work of the railway, and how they changed in appearance. Enjoy.

Chapter 2: The 50 Class 50s

50 001 Dreadnought

50 001 had been a Western Region engine for just over three years by the end of the long hot summer of 1976. Looking a tad weather beaten and displaying its identity in the now redundant head code reporting box, 50 001 makes light work of the 1 in 57 gradient on the approach to the 565 yard Treverrin Tunnel, the summit of the line between Lostwithiel and Par, with that time honoured train 1V76 0920 Liverpool Lime Street - Penzance. I had actually come here for D1068 Western Reliance after photographing it reflected in the high tide at Golant, knowing this would immediately follow 50 001 up the bank with its Fowey Docks - St Blazey clay empties. What would we now give for variety such as this?
27 September 1976

A decade later same engine, same train. It did haul other services!! Refurbished, now in the popular large logo livery, and named Dreadnought on 10 April 1978, 50 001 rolls in to Newton Abbot crossing the River Teign with a fine array of signals on display with 1V76 0920 Liverpool Lime Street-Penzance. Noticeable as it will be henceforth in many of the views following refurbishment to come in this chapter, after headcodes were no longer used the lights that illuminated the blinds were reused as marker lights. Some The view taken from Newton Abbot East Signal Box, the second largest frame on the GWR with 206 levers, a couple months before the area passed to the control of the Exeter Panel Signal Box. To the right of the train is Hackney Yard, to the left Newton Abbot Race Course. 06 March 1987

50 002 Superb

A smart looking 50 002 in its early career with the WR: In glorious late winter sunshine, 50 002 heads east over the 10 arch, 738 feet in length and a maximum height of 70 feet Coombe St Stephen Viaduct between Burngullow and Grampound Road with charge of 1A19 1100 Penzance - London Paddington, the Up Cornish Riviera Express. In the foreground are the piers of the original Brunel timber viaduct replaced by the present structure 11 July 1886. Sadly due to growth this fine view is no longer available. 14 February 1976

Only five years have elapsed since the picture above, but the engine does not look quite the same after refurbishment with the marker lights in the former head code box, and proudly on display its Superb nameplate gained 21 March 1978 adorned with the ship's crest above. With a fine rake of Mark 1 stock in tow, 50 002 brings the sort of train that characterised summer Saturdays in the West of England namely 1B78 0755 Worcester Shrub Hill - Paignton around the curve from Dawlish Warren past Langstone Rock on to the sea wall. The engine survives in preservation and indeed in Devon, currently now under restoration on the South Devon Railway. 20 June 1981

50 003 Temeraire

Looking spick and spam having settled in well on its transfer from the LMR, a yet to be named 50 003 heads through the cutting at Dobwalls, just west of Liskeard, with 1B25 0930 Paddington-Penzance. I suspect at the time, some choice words were uttered this was not one of increasingly rare class 52s. The moral is always take the picture. 19 June 1976

From 50 003 in its original condition above, here we catch up with the engine after refurbishment and naming (9 May 1978) looking resplendent on a humble duty at Langham, just to the west of Ivybridge, with an Up Permanent Way Special. The consist is a train empty and rather smart grampus wagons, these used to convey ballast. By this time HST's had made serious inroads to many of the prime passenger turns releasing class 50s for more mundane work. Sadly recent excessive growth has made this view almost impossible. 10 May 1983

50 004 St Vincent

Looking very smart in its original condition, 50 004 hurries west past Gissons near Exminster with 1B55 1230 London Paddington-Paignton. The train has just passed under the new viaduct to carry the M5 motorway still under construction at this time. 25 August 1976

Six years on from the picture above; now named St Vincent 9 May 1978, refurbished and looking very smart in its large logo livery, 50 004 in charge of 1V76 0920 Liverpool Lime Street-Penzance. A truly magnificent sight upon St Germans Viaduct over the River Tiddy, a tributary of the River Lynher which in turn is a tributary of the River Tamar. This graceful 13 arch structure was brought in to use 19 May 1908 on completion of the deviation of the original route from Wearde (immediately to the west of Saltash). Sadly this view is now impossible due to growth, but a satisfactory picture can be taken at a lower level on St Germans Quay. 23 March 1982

50 005 Collingwood

Not looking quite so spick and span, 50 005 is about to conquer Dainton Bank and enter the tunnel of the same name with a mixture of Mark 1 and II stock 1A69 0835 Penzance-London Paddington. Sadly an angle no longer available due to growth and access problems, the cross over between up and down lines is also no more. 31 July 1976

A picture that will conjure happy memories for many who remember and enjoyed those halcyon summer Saturdays of the class on the Par-Newquay Branch. With just one more weekend of locomotive hauled services over the line remaining, 50 005 (named on 05 April 1978) waits patiently at the line's terminus to depart with 1A51 1040 to London Paddington, the train, other than a rail blue BG, formed of Network Southeast stock. This was still the era of British Rail where everything despite its sponsored branding was very much common user. I wonder how many of the returning holiday makers would be sat in those coaches on the Monday morning when the rigours of the daily commute to the capital would resume. 26 September 1987

50 006 Neptune

In lovely early spring sun shine, 50 006 with 1A45 1235 Penzance-London Paddington is seen passing through the second Burngullow station, closed 14 September 1931, the original of 1863 was to the east of the road bridge on which I am standing. The then still extant signal box on the former down platform closed on 5 October 1986 when the track westward to Probus was singled, and after causing a bottle neck was re-doubled at great expense in November 2004. Behind the train is a small building that was on the former up platform.

The line curving away to the right is the still in use freight only branch to Parkandillack, Crugwallins and Drinnick Mill.

After a period of spectacular growth, this location was cut back in 2020 and a reasonable picture can again be taken here. 14 February 1976.

Neptune, the first Class 50 to be refurbished, bore its name from 25 September 1979. Now in large logo livery, 50 006 is captured from the top of the 1092 yard Whiteball Tunnel leading 1S71 0730 Penzance-Glasgow Central and Aberdeen as it is about to leave Devon and emerge in Somerset. Within a year the scene would change here with the disappearance of the semaphore signalling as the area was taken over by the Exeter Panel and the local signal box (barely visible to the left at the extreme end of the train just in front of the road bridge) being closed on 9 March 1986. At the same time the siding was also removed leaving just plain track here. 27 July 1985.

50 007 Hercules

4B10 1553 Bristol Temple Meads-Plymouth vans was a time honoured train of the diesel era, and one to watch out for as motive power could range from a class 25 upwards. It was also notable as a suitable way of returning engines not in the best of health back to Laira Depot for some much needed TLC. In a scene which reflects the ethos of the Western Region where at this time little had changed, 50 007 in fine fettle heads west through Tiverton Junction with 4B10 with a fine selection of vans in tow. Note beyond the footbridge the impressive line up of lower quadrant signals in the Up direction. 11 September 1976

50 007 after naming (6 April 1978) and subsequent refurbishment being one of the last to emerge in this form from Doncaster Works in February 1983 looking very smart in large logo livery heading east over the River Plym at Crabtree with 1Z15 0815 Penzance-Paddington Relief 23 September 1983. The view is from the footbridge to Saltram Park immediately in front of the A38 Plympton by Pass. It can still be obtained, but with a much changed back ground with Alfred Bell's Warehouse and the Bass Charrington Brewery Depot behind the train all eradicated to make way for Sainsbury's Super Market, and even this can hardly be seen in the veritable jungle that now exists here. The engine would 5 months after this view be plucked from obscurity to become the first celebrity member of the class. All is revealed on the following page. September 1983

50 007 Sir Edward Elgar

50 007 was one of the only two of the class to be renamed. After just a year in the large logo livery, it was repainted into Brunswick Green livery and unveiled at Laira Depot as Sir Edward Elgar in February 1984 to mark the 50th anniversary of the death of the celebrated composer. Why was this engine chosen for such an honour one may ask? Take the numbers and leave out a 0 and rearrange them and one comes to GWR Castle 7005 Sir Edward Elgar. The latter spent most of its working life allocated to Worcester Shed (85A) which was of course very much in the part of the country associated with Sir Edward. There is another similarity between the two engines as the composer's name was the second name 7005 carried, being renamed from Lamphey Castle in August 1957. 9 April 1987 witnessed one of the few occasions a member of the class worked a Royal Train. This is 50 007 arriving at Plymouth, and passing the former Royal Eye Infirmary with a selection of cars of the period parked up in Ermington Terrace. On this occasion the very special train was conveying the then HRH Prince Charles.

I was in charge of the Ticket Office at Plymouth Station that morning, we had to lock up and clear the concourse about 20 minutes before the Royal party arrived. The temptation to nip up the road for a picture was too tempting. Running back to the station I found a large queue outside, which I jostled to get past only to find a policeman telling me entry was barred until the Ticket Office was reopened; my reply was "I got the keys in my pocket!!".

The reaction on hearing about this escapade from the then Station Manager was a request for both a print and duplicate slide of the event. The locomotive has the unique distinction of being withdrawn twice from service by BR, firstly on 16 September 1991 only to be returned to traffic early the following year for rail tour duties including the last one of all, the 50 Terminator of 26 March 1994 to then immediately be withdrawn again. Its career since has seen change of ownership more than once. From the world of the Heritage Railway it is currently owned by the class 50 Alliance who in agreement with GB Railfeight has seen it repainted in the distinctive GBRF livery and certified for main line work, in effect putting the engine back into capital stock.

Following its transformation to Sir Edward Elgar, 50 007's celebrity meant the locomotive was frequently in demand for depot open days. In later years it became Laira's mascot, after all it is not every depot that had a ready made icon. When it was transferred into the NSSA Waterloo Exeter pool in 1989 Chris Green then Director Network South East instructed that it be repainted into the new striking NSE livery. However, Geoff Hudson AFM at Laira had other views, and on explaining its significance as the depot flagship, Mr. Green conceded, and the deal was done that it would remain in the green livery, provided it was kept in ex works condition. The early 1990's was an unsettling period in the rail Industry with the lead up to rail privatisation, and the flagship of Laira was to provide an important role in demonstrating the pride and achievements of the depot, and as such, survived up to the end of British Rail.

During its period in green, number seven received no less than 5 repaints in 4 different shades of green. This photo shows the third iteration following an F Exam in April 1987 and the shade is believed to be Lotus Cortina Green. Yes, the green stripe down the side of white MKII Lotus Cortinas!

50 008 Thunderer

50 008 stands at Penzance waiting to take 2C89 1815 to Plymouth, unusually for the Up evening local service it bears The Wedding Belle headboard borrowed by BR from the Dart Valley Railway. The occasion a very special one in my life, the train was returning my new wife Audrey and I along with relatives and guests from our wedding at St Buryan earlier in the day. This was very much a railway themed wedding. Entrance to the church including officiating clergy, groom, best man and the bride and her father, was only by a specially printed St Buryan Edmondson card Church platform ticket which had to be clipped by the ushers. After the ceremony, the happy couple staged a photographic walk past for the gallery of photographers more used to moving trains. One of the most memorable moments was at the reception (in the village hall) when one of the officiating clergy the then Bishop of Truro, the late Peter Mumford, stood up and proclaimed "this is the only wedding I have ever attended where the main toast was to the Great Western Railway, and I shall remember this for some sermons!" Just to add a bit more spice to the occasion, before carrying my new wife over the threshold I carried her to the footplate of 50 008 for a ride up front from St Austell to Par. Start as you mean to go on!! 11 May 1985

G R O O M
0446
Mr. B. M. Mills & Miss A. E. Fryer
ST. BURYAN
CHURCH TICKET
Available on day of issue only
Issued to Commemorate the Marriage of
Bernard and Audrey on
SATURDAY 11th MAY 1985
0446
B R I D E

The Church Platform Ticket

Near to the end of its career with BR 50 008 (named on 1 September 1978) was painted in what became known as "Laira Blue" livery, created by Laira Depot primarily to identify locomotives transferred nominally to Departmental use. 50 008 very much at this time also an engine in demand for rail tours, and was the only one of these so treated to receive small numbers.

Looking radiant, the engine poses at Laira on an inspection special, on the line through the depot known as "The Speedway" which remains as the eastern access route to and from Mount Gould Junction. Note the crest above the name plate presented by HMS Thunderer, not then a ship but the Royal Navy Engineering College at Manadon in the leafy Plymouth suburb of Crownhill. The engine survives in preservation, RNEC Manadon closed in 1995 and just like many a former railway installation no trace of it remains, the site now a housing estate. 04 June 1991

50 009 Conqueror

A day and a month after being named Conqueror 50 009 heads west out of Plymouth Station over Cornwall Loop Viaduct with 2B26 1745 Plymouth-Penzance. The view is from Victoria Park. In front of the train are the piers of Stonehouse Pool Viaduct demolished in the autumn of 1965, this was the original route of the Cornwall Railway of May 1859 westward out of Plymouth Millbay. This scene was getting difficult to capture even then. Now it is impossible due to a combination of growth and since the Millennium, what is best described as a piece of sausage shaped iron work supposed to represent Brunel was placed on the three piers of the redundant viaduct to the right of the picture. 09 June 1978

A view now not available from Par Green due to further expansion of the clay processing industry in these parts, 50 009 climbs the bank away from Par station up the 1 in 56 grade with 1V10 0920 Brighton-Penzance. This is very much in the heart of China Clay country; to the left of the train behind the coaches note the six pan kilns of Par Moor easily marked by the chimney above the second coach adjacent to the A3082 road. In latter days some were used for storage until 2003 and have since been eradicated without trace. Behind the train the Cambrian Bulk Clay Store and plant in Par Harbour are busy processing the Cornish white gold, some traffic from here to the Potteries is still carried by rail. 13 August 1983

50 010 Monarch

50 010 on the edge of Dartmoor at Palace Lane on 31 May 1978, here the up line side fence to the right of the train also marks the boundary of the National Park at this point. Named Monarch only six weeks previous 16 March 1978, 50 010 is motoring well with 1A13 1224 Penzance-Paddington. The location is now spoilt for photography by the appearance of a large Network Rail mast to the immediate left of the scene, and the adjoining field becoming a housing development. Rutt Lane over bridge is to the rear of the train and it is just immediate to the other side of it the second Ivybridge station was opened on 15 July 1994.

A classic Newquay Branch scene to stir the memory; 50 010 now in the large logo livery breasts the 515 feet above sea level summit of the line at Tregoss, having just crossed the then A30 road, with 1A75 1035 Newquay-Paddington. This view is now a very much a head on one due to growth, another change the main A30 road was diverted to a dual carriageway just to the north in late 2007, the bridge and road still in use at this point for local traffic. 18 June 1983

50 011 Centurion

Treesmill is about a mile east of Par on the climb to Treverrin Tunnel with 50 011 on the 1038 Newquay-Paddington. In those days before mobile phones and gen, an insight in to how much stamina and endurance the younger body could stand, and what could be achieved on a day out in Cornwall and give an impression of the class 50s seen that day. The early start to leave Plymouth was not a problem as I had just finished a week of night shift in the ticket office at Plymouth Station, leaving there at 06.00, on the way collecting my pals Robert "Cyril" Lord and the late Maurice Truscott. Call number one to intercept 50 022 coming off the Royal Albert Bridge with 0457 Penzance-Paddington. Then a fast ride to the Falmouth Branch for 50 018 on 2B70 0814 Truro-Falmouth at Penwithers and chased to Penryn, the return 1A59 0910 Falmouth-Paddington on Carnon Downs Viaduct, and beat it to Grampound Road. There followed a swift ride over to the Newquay Branch for 50 011 at Tregoss, Luxulyan, St Blazey and then a very fast short run to get the fourth picture here at Treesmill. Back to Par station for 50 044 on the 0800 Bristol TM-Penzance and 50 047 on the 1050 Penzance-Paddington, the last day the Up "Cornish Riviera" would be locomotive hauled. More action on the Newquay Branch with 50 016 at Bugle on the 0948 from Paddington. before calling in at Lostwithiel for 50 048 on the 1600 St Austell-Kensington Olympia Motorail on the way to the fish and chip shop at Saltash. Leave the car at home and over to the Archer Inn for some Draught Bass from the barrel and at midnight a very welcome bed!! Notice parked up in a convenient lay by in the country lane by the bridge at Treesmill my little Ford 100E car, a celebrity with its replica Western Empire nameplates, a vehicle that did a lot of miles and gave great service in the noble cause of railway photography. 04 August 1979

50 011 has the unfortunate distinction of being the first of the class to be withdrawn by BR on 24 February 1987, it would take another seven years to eradicate the class. I remember seeing it on its last working at Plymouth as I had been told 50 011 would head 4M05 Penance-Crewe vans from there through to its destination for withdrawal and storage at Crewe Works. In happier times, refurbished and bearing its Centurion name from 17 September 1979, 50 011 passes Laira Junction to the east of Plymouth working 1Z15 08.15 Penzance-Paddington Relief on 24 October 1983, the consist a neat rake of Mark 1 stock which was becoming a rare sight by this time on London workings. Does anyone remember relief trains? Such a term does not appear to exist in the privatised railway of today. 24 October 1983

50 012 at speed with the diverted away from the Berks and Hants route to run via Bristol 1A29 0915 Penzance-London Paddington passing through the site of the former Box Mill Lane Halt. The little station had a comparatively short life of only 35 years opening 31 March 1930 and a victim of the Beeching cuts closing 4 January 1965. It was actually more convenient for the village of Box whose station dating from the Brunel era was situated on the Bristol side of Middle Hill Tunnel seen in the back ground. This as we shall see in some of the forthcoming pages was a very productive trip on a lovely summer Sunday, with the bonus of the West of England trains as well as the normal service on the Bristol main line. I must admit the spur for the long drive and day out from Plymouth around the Box area was to see what class 52 Western's would appear in the lens. 15 August 1976

Moving on just a couple of years, not long after being named on 3 April 1978, 50 012 proudly displaying her Benbow name plate (brought back memories of D805 Benbow) at speed on the curve at Powderham in charge of 1M74 1358 Penzance-Birmingham New Street, a train instantly recognisable with the buffet always at the rear end. 21 July 1978

50 013 Agincourt

Recalling the comparatively short heyday of the class in command of the Bristol expresses, 50 013 on the gentle curve in the cutting at Corsham on the approach to Box Tunnel with 1B32 1345 London Paddington-Bristol Temple Meads on 15 August 1976. Note, the train a neat formation of mostly air conditioned Mark 11 air conditioned stock, the colours reflect the drought of the summer of '76.

From the brash colours of a long hot summer to the mellow tints of winter; named Agincourt 19 April 1978, 50 013 1V76 0920 Liverpool Lime Street-Penzance by now formed of air conditioned stock and a Mark 1 buffet car (remember them?) on the approach to Dawlish Warren station. Note the fine array of WR semaphore signals. 12 February 1983.

50 014 Warspite

On the flat lands to the south of Exeter, 50 014 heads east at Gissons passing the Exminster Up Starting Signal 7B64 1255 Plymouth Friary-Exeter Riverside Goods, the traditional sort of train now long consigned to the history book with its mix of traffic and wagons. Freight from Plymouth Friary is also now a memory. The locomotive would receive its Warspite name 30 May 1978, (my 31st birthday!!) 25 August 1976

A long standing and favourite train which still lives in the hearts of many was 1M74 1358 Penzance-Birmingham New Street, which throughout the diesel era was renowned for its more often than not double heading through Cornwall. There was no operational need for this train to be double headed, it was just a convenient way of working a spare engine back from Penzance to Plymouth, and any combination was possible. For class 50 devotees this is the train at its best as 50 014+50 002 provide super motive power for the service at Tresulgan, the viaduct of that name in the background, After years of growth, this angle became available once again after severe cutting back by Network Rail in the spring of 2020, the afternoon entertainment here these days not so entertaining. 15 September 1980.

50 015 Valiant

Still in its original livery and about six weeks on from receiving its name (21 April 1978), excuse the pun but I could not resist the temptation, a Valiant effort against the early evening sun. Other than the engine on the front, little had changed in this scene from steam days with the mix of milk tanks and assorted parcels vehicles to the line of telegraph poles. The latter were until fairly recent times a feature of the line side, and often the bane of many a railway photographer. 50 015 leans in to the curve on the descent to Par at Carlyon Bay with 6019 16.22 St Erth-Clapham Junction Yard with on the drawbar a impressive load of milk tanks and parcels vehicles. The milk traffic would cease in a year and two months on from this picture. 03 June 1978

Towards the end of its career, 50 015 was painted in what was known as Dutch livery to indicate its nominal transfer to the Engineering Department, which did not stop the engine appearing on rail tours. With the sun playing fast and loose, 50 015+Saloon DB999506 form 1Z01 0915 Plymouth-Newquay Inspection Special on the approach to Lostwithiel passing the fine down home two arm bracket signal. The engine survives in preservation. 30 October 1991

50 016 Barham

More Newquay Branch action at a location rarely photographed. 50 016 is delighting the class 50 followers with a fine plume of exhaust coming over Molinnis Level Crossing on the 1 in 67 grade from Goonbarrow Junction. It is about to pass through Bugle station with 1B48 09.48 Paddington-Newquay. The rusting rails in the foreground were once part of the double track section of the branch from Goonbarrow Junction to Bugle. This was abolished from 29 November 1964 when it became two parallel single lines, the inner one being the Carbis Wharf only freight only branch. This saw its last train on 25 August 1989. Today just the rails to Newquay remain in situ. 04 August 1979

As the days lengthen 50 016 named Barham 3 April 1978 looking very smart in large logo livery nears the top of St Germans Bank at Treskelly Farm working 2C74 1557 Plymouth-Penzance; the country side of South East Cornwall at its best in the late afternoon spring sunshine. Working 4 or 5 coach Cornish local trains perhaps a menial task for the class, but far better for the passengers who in just a couple of months would find themselves relegated to the delights of the overcrowded 2 car Sprinter which would take over these workings. 01 March 1988

50 017 Royal Oak

Not far to go to journey's end. Looking smart in original livery, 50 017 with a fine plume of exhaust and some wonderful sound pulling away from St Erth up the 1 in 67 past Canon's Town with 1B25 0930 Paddington-Penzance. 29 May 1976

I like the name Royal Oak for a number of reasons. Firstly because the name was bestowed on class 43 Warship D842, an engine I have fond memories of. Secondly, Royal Oak is a fine real ale brewed by Hanlon's Brewery in Exeter, a pint I have sampled on many an occasion. Thirdly, it is the name of an excellent local hostelry, the much recommended Royal Oak in the nearby village to me of Meavy, unique in being owned by the local parish council. Last and lot least, 50 017 Royal Oak was based for a while the Plym Valley Railway which I founded and was superbly restored there to its NSE livery, making it the largest and heaviest engine to be seen on the former GWR Tavistock and Launceston Branch. Four years to the day since it had been awarded its Royal Oak nameplate, still in original livery but looking a bit worse for wear, 50 017 stands with 50 022 at rest on Laira Depot. 24 April 1982

Rare territory for one of these in normal service gracing a Cornish branch line that rarely saw locomotive hauled trains in the diesel era to boot, the St Ives Branch. Top and tailing with 37 299, 50 018 named Resolution 6 April 1978 cautiously approaching St Erth from St Ives with the annual weed killing train. It is coming off the long curve from Lelant Saltings Park and Ride Station which can be picked out in the right hand back ground. A bit of a rushed shot as photography further up the branch met problems, and the chasing party caught in heavy traffic arriving here as the train entered the cutting, hardly time to even get out of the car! 29 April 1982

It is in the leafy suburbs of Surrey as the early signs of spring are showing. A lovely portrait of 50 018 working 1V15 1515 London Waterloo-Exeter St David's at speed as seen from the King Charles road bridge at Surbiton on 21 March 1991. Surbiton for ever a place in television folk lore as this was the setting for that classic BBC comedy series "The Good Life". Sadly it would not be a good life for 50 018 withdrawn just four months and a day after this picture was taken, and my local spies tell me this splendid view is now spoilt by growth. 21 March 1991

50 019 Ramillies

One of the joys of railway photography is recording the ever changing seasons. In low winter light with its lengthy shadows, 50 019 named Ramillies six months earlier on 18 April 1978 but still in original livery approaching Lostwithiel 1B44 0935 London Paddington-Penzance The view was taken from the main A390 road bridge on the eastern end of the town. Top left Restormal Castle set atop its hill keeps watch over the proceedings, to the right the cows graze on in blissful ignorance of the passing trains. 13 November 1978

In 1989 50 019 was transferred from NSE to Departmental use. The boss of NSE Chris Green was not impressed to see a scruffy locomotive bearing his colours now on more menial duties such as ballast trains, and requested it be repainted. Laira Depot duly obliged but wanted something neutral so they repainted it back into BR blue with black window surrounds, a grey roof and over size NSE numbers because normal size transfers were no longer available. A rather well worn looking 50 019 waits to leave Platform 7 at Plymouth with the fairly lengthy 2156 Leeds-Penzance parcels. 12 December 1989

50 020 Revenge

With the head code box thinking the locomotive is Trematon Castle, 50 020 speeds west at Westcott, between Cullompton and Hele & Bradnich, alongside the brand new, and for a summer Saturday remarkably empty, M5 Motorway with 1B09 0930 Paddington-Paignton. 12 September 1976

Through wide open spaces where just a couple of decades earlier this would have been a scene of hustle and bustle. 50 020; named on 7 July 1978 and looking very smart in the large logo livery leans into the curve away from the former Gwinear Road station (its remains just visible towards the rear of the train on the west side of the level crossing) at Polmenor Downs in charge of 1E36 1117 Penzance-Newcastle. Hard to believe the lush field in the foreground was the site of Gwinear Road Marshalling Yard which served freight traffic to and from the former Helston Branch, between where I am standing and the train there would have been eight lines of rail. The yard was taken out of use 31 October 1965, and as can be seen not a trace of it remains. The Helston Branch went off behind the house to the left. The hamlet of Gwinear Road taking its name from the former station closed 5 October 1964 is behind the engine to the right of the trees, the pasties in its farm shop much recommended. 24 June 1986

50 021 Rodney

From Gwinear Road a short trip along the main line further west to Marazion whose station also saw its last passengers on 3 October 1964. 50 021 is opening up after passing through the remains of the station with 1B11 0830 Cardiff Central-Penzance. Towards the rear of the train to the right of the building on the former down platform can be observed the line of six Pullman coaches, which since the early 1960's had been used as camping coaches. Their position on the edge of Mounts Bay open to the ravages of the Atlantic weather did in the end cause their downfall as they rotted away. By the summer of 2003 they had lost the battle with the elements and had all gone, their loss a sad one as they had become a bit of a local landmark. 29 May 1976

From the far west of Cornwall to the eastern extremity of Devon, 50 021 seven years later having been named on 31 July 1978 has charge of 1M85 0740 Penzance-Liverpool Lime Street on the final ascent to the summit at Whiteball .This was a classic location especially with the lower quadrant signals. After a period of being blocked by growth, as a result of clearance work by Network Rail this scene can be repeated, but without the down loop line and the signals, lost when the Exeter Panel Signal Box took over in these parts over the weekend of 27 April 1986. The locomotive survives in preservation. 11 June 1983

50 022 Anson

When the railway was neat and tidy and the view not ruined by growth, and when traditional signalling and the infrastructure of the steam age were still much in evidence .50 022 is about to conquer Dainton Bank from the east and enter Dainton Tunnel with the Sunday 1B16 0808 London Paddington-Plymouth (via Yeovil). This is a much changed scene here today, and this angle is now impossible due to a combination of growth, an electric fence and access problems. 16 October 1977

Near to the end of the descent of Filton Bank at Narroways Hill Junction to the north of Bristol is 50 022. The engine was named on 20 April 1978, and is looking fine in the large logo livery. This is 1V76 0920 Liverpool Lime Street-Penzance, a train already vary familiar in these pages. It is passing the abutments of the bridge which once carried the direct link from the Avonmouth line, Ashley Hill Junction to the former Midland Main Line at Kingswood Junction. This became disused as of 14 June 1965. The quadruple track from Filton Junction to Dr Days Junction was reduced to double track in the autumn of 1984.. As services from Bristol to both South Wales and the Birmingham route were enhanced over the years, two lines of rail caused a bottleneck, the quadruple track over the whole length was restored at great expense by Network Rail over the autumn of 2018. 27 May 1983

Passing the 283¾ mile post from Paddington measured in these parts via Bristol Temple Meads, Weston-Super-Mare avoiding line and Millbay is 50 023 with a lengthy 2B26 1745 Plymouth-Penzance. It is a warm and sultry evening as 50 023 named only 3 weeks previous climbs the 1 in 57 grade past Mertham just around the corner from Par Green Farm towards Carlyon Bay, note the industry of Par Harbour in the far background. 03 June 1978

Five years later almost to the week and again bathed in the evening sunshine, 50 023 which had been named on 17 May 1978 takes the Cotswolds route to Oxford at Norton Junction, just to the south of Worcester, with 1A47 the Sunday 1815 Hereford-London Paddington 12 June 1983

50 024 Vanguard

It fits nicely!! In the bleak mid winter as Christmas has passed and the New Year beckons, a seasonal touch as the morning frost has yet to thaw, 50 024 emerging from 291 yards Dainton Tunnel with 1B15 0830 London Paddington-Plymouth. Dainton Bank in steam days was a stern test for crews, the diesels coped better, although depending on the load the further up the bank, the slower they became. I certainly heard 50 024 before it entered the tunnel. Knowing the signal-man in the adjacent, and then still staffed, signal box helped gain access to this private location. 29 December 1976

With 16 bogies on the drawbar, 50 024 (named 15 May 1978) has its work cut out with the fully laden 1M01 Sunday 1620 St Austell-Kensington Olympia Motorail on the approach to Dawlish Warren. The Motorail brand, established in 1966, saw its heyday in the 1970's. Sadly this was seasonal traffic and as the Motorway system expanded, cars became more reliable, fares increased, patronage waned and services finally finished in 1995. By the time of this picture this train only ran to capacity at weekends and finished at the close of 1981 season, being replaced for three years by a separate train for the cars only from Paddington; the occupants travelling on a normal service train. With inevitable gap between the arrival of the car and the people, or vice versa, at their destination hastened the final decline of the service three years later.

The beauty of Box: from the top of Middle Hill Tunnel a panorama of rural Wiltshire as 50 025 races along the straight out of Brunel's Box Tunnel through the site of the former Box Mill Lane with 1B24 1430 London Paddington-Paignton on 22 August 1976. This was a return visit to Box a week after a very successful visit to the area to try and get this view with a class 52 as the 1430 from Paddington on a Sunday that summer had been almost a solid Western turn. I must admit at the time some unprintable words were uttered, in retrospect I am glad I did press the shutter as it presented me with a now unrepeatable scene of a classic locomotive in a classic setting. The moral of the story is always take the picture as it is tomorrow's history, and I did get a class 52 here the following month.

Other than for visits for refurbishment and other major work, a pair of class 50s as captured here well off their normal beaten tracks at Doncaster. 50 025 (named 6 June 1978) + 50 040 lead the stock for the 1Z44 0550 Plymouth-Doncaster Dart rail tour. I find it amazing that with the rare sight of pair of class 50s gracing the East Coast Main Line, the platforms are devoid of onlookers and photographers jostling for position. 26 July 1981

50 026 Indomitable

This was the way to travel, class 50 and Mark 1 stock – a classic combination indeed. Named on 29 March 1978, after refurbishment and bearing the large logo livery 50 026 heads 1B58 1010 London Paddington-Paignton heading west at Norton Fitzwarren. The rusty rails to the left of the train are the only reminder of past glories here such as the limit of the quadruple track from Taunton, the former Up relief existing as the main line connection to the former Minehead branch which by this time had become the West Somerset Railway, and this is still in use today. In the back ground the busy yard and works of the Taunton Cider Company. 11 June 1983

50 026 is as far as my lens is concerned the most camera shy of the class, and I have few pictures of it in pre Network Southeast days, thus a further reminder of the joys of the large logo 50 026 and Mark I stock. This time with '26 heading west through the fields of Southeast Cornwall at Bethany as spring 1983 has sprung with 1V10 0920 Brighton-Penzance. Sadly the notebook for this period is mislaid so precise date isn't known. The location is between St Germans and Menheniot from an over bridge on a quiet country road. Sadly one needs a long lens to get a decent picture here these days to avoid the foreground growth.

50 027 Lion

A striking almost head on view as 50 027 in charge of 1A37 0855 Paignton-London Paddington makes headway around the S bend at Hollacombe, the over bridge in the back ground marking the summit of the climb of Preston Bank out of Paignton. The train is passing what was to the right the site of Torquay Gas Works. A difficult location to photograph and very much an early morning (by my standards!) one, growth even then necessitating use of a 135mm telephoto lens to overcome the foreground growth. The area now has its own private jungle. 02 June 1984

There was nothing quite like the Newquay summer Saturday morning line up at Cornwall's longest platforms. To the left is the stock of the 1005 to Manchester with 47 7615 out of sight on the front. Taking centre stage is 50 027 which had received its Lion nameplate on 17 April 1978 waiting with its 14 coach 1E65 0925 Newquay-Newcastle. To the right 50 005 waits its turn with 1A51 1040 to London Paddington All this activity is seen on the penultimate weekend of locomotive hauled trains to Newquay and two weeks before the closure of the signal box and the brutal rationalisation of the track capacity leaving just a single line into Platform 2 where we see 50 005, and not even a run round loop was provided. How the mighty fall. 26 September 1987

50 028 Tiger

Looking very spick and span in its original livery, 50 028 is slowing down for the Exeter St David's stop with 1A45 1220 Penzance-London Paddington . The train is running beside the very convenient footpath for railway photography at Flower Pots, Exwick. The harsh colours of the long hot summer of '76 are evident, the angle now affected by growth. 03 July 1976

The sun does not always shine on the Dawlish sea wall. In sultry summer conditions, 50 028 with another classic summer Saturday train namely 1V63 0803 Wolverhampton-Penzance. The engine looking very smart with its nameplate applied 10 May 1978 makes her way from Dawlish Warren towards Rockstone Bridge, the landmark which is Langstone Rock prominent in the right hand background. 20 June 1981

50 029 Renown

50 029 passing Swindon Works to the left of the picture with the diverted 1B83 Sunday 1530 London Paddington-Penzance. Back then the rules were not so strict and standing by the main line on the tracks officially part of the works complex was not a problem. Probably need an Act of Parliament now for such goings on. Behind 50 029 is St Mark's Church, built 1843-45 with the assistance of the GWR to serve the spiritual needs of the workers who had swelled the population of the town on the establishment of the works here, designed and inspired by both Sir Daniel Gooch and I K Brunel. So this is where God's Wonderful Railway came to pay homage to its creator. 26 June 1977

One has to stop and wonder, did Newton Abbot really once look like this? 50 029 named on 26 October 1978 heads west away from the station with 1V10 09.13 Brighton-Penzance. So much has changed here, to the right where we see the David & Charles stock is now a supermarket, the shed and yard now an industrial estate. The track layout was severely butchered in May 1987 when the Exeter Panel took over leaving only three lines of rail and platforms in use, the down main where see 50 029 and those either side. To the left of Newton Abbot West Signal Box this area is now a car park. Growth also now impairs the view and through trains from the south coast to the south west are also now history. At least the engine survives in preservation. 29 September 1986

50 030 Repulse

The first class 50 working out of Waterloo was on 5 November 1979, the class being fully diagrammed for all workings on the route came with the May 1980 timetable. A study of 50 030 Repulse in original livery bearing the nameplate it obtained 10 April 1978 passing Pinhoe with 1O12 1143 Exeter St David's-Waterloo on 24 May 1980 very early in the era of class 50 domination of the Waterloo-Exeter route. To the right the battered gate once protected entry into the wartime sidings laid circa 1943 for a Ministry of Food frozen store, the large brick building behind the Nissen huts. All this has been swept away to make room for a housing development and growth now impairs the view.

A busy day to be in Cheltenham was Gold Cup Day. One could avoid the crowds by adjourning to the Cloddymore footbridge just to the west of the station and enjoy views like this. 50 030 slowing for the Cheltenham Spa stop with the lengthy 1S71 0730 Penzance-Glasgow/Aberdeen. One has to wonder how a 4 or 5 coach Voyager can really replace a long busy train like this. The angle is still worth a look, and the down loop to the left survives in use. 14 March 1985

50 031 Hood

The former Tiverton Junction station where 50 031 looking rather grubby in its original livery and thinking it is 5031 Totnes Castle arrives with 1B73 1430 London Paddington-Paignton. Sadly by this time the station was a junction no more. Look behind the joining of the first and second coaches of the train to see the course of the former Hemyock Branch curving away to the right, closed 31 October 1975 on cessation of the milk traffic. It did not take long to eradicate the branch here to make way for a very recently opened stretch of the M5. The former ESSO oil terminal is opposite to the left, and there is a fine array of signals. Other than the four tracks, as the loops remain in use, there is really nothing now to see, and with the footbridge no more, there is no way of updating this picture. 11 September 1976

50 031 slowly rounds the curve at St Dennis Junction with 1V39 0723 Manchester Piccadilly-Newquay. The signal box had closed the previous year after the summer season was over, and no time was wasted in demolishing it and removing what was a useful crossing loop. Here in the Cornish wilderness where Goss Moor meets the Fal Valley, three lines of rail route once met. The Retew Branch to Meledor Mill closed on 3 April 1982 and went off to the far right on a 1 in 40 gradient. The rails to the left of the latter served the tip and were the remains of the line from Parkandillack (closed 6 February 1966). This view was taken from the former A30 road bridge, and needless to say it is a totally unrecognisable scene here today. Other than the single line to Newquay all other lines have been eradicated, the A30 Indian Queens by Pass bridge of October 1995 now dominates the scene blocking any decent view. 01 August 1987

50 032 Courageous

Overshadowed by its neighbour the more famous Box Tunnel half a mile up the line,50 032 emerges from the 198 yard Box Middle Tunnel with 1B22 1120 London Paddington-Weston-Super-Mare . It may not have the fame of its larger neighbour by having its western portal on display from the Bath Road, but its portal is just as ornate and an architectural gem in its own right, Brunel sparing nothing to embellish his handy work. I am told growth and fencing now spoils this classic view. 22 August 1976

A West London panorama from a convenient block of flats with 50 032 (named on 7 July 1978) approaching Subway Junction with 1A75 0700 Hereford-Paddington. In the foreground are the tracks of the Hammersmith & City line once jointly owned by the Metropolitan and the GWR Companies, to the right the elevated Westway opened in 1970 carrying the congested A40 road. Towards the rear of the train a couple of Route Master buses cross the tracks by the exterior of Westbourne Park station. A much changed scene here now with electrification of the main line, and development of the land beneath Westway. 50 032 did not escape the cutter's torch unlike the last ship to bear the name, the Churchill class nuclear submarine HMS Courageous, decommissioned 10 April 1992, has been preserved (with the dangerous bits removed) in the former South Yard of Devonport Dockyard, and is planned to be the centre piece of a Cold War Museum. 01 March 1979

50 033 Glorious

The winter shadows fall over Saltash Station where I arrived in October 1965 to learn my trade as a Ticket Office Clerk – a threatened species in the current climate – as 50 033 slowly passes through as there is a 15 mph restriction for all trains crossing the Royal Albert Bridge with 1A47 0838 Penzance-Paddington. The station became unstaffed in June 1972. The footbridge was demolished 9 April 1989. Above it dominating the scene was Saltash Baptist Church which was burnt down on Christmas Day 1987, replaced by a new building in March 1992 but any view of it from the station is now obscured by the trees. At least the station is still open for passengers. 25 November 1978

Two for the price of one: A timetable change had brought about 1M74 starting from Plymouth at 1618. It was replaced in Cornwall by the 1345 Penzance-Plymouth local service which also brought up the vans. Fortunately this still offered on some days double heading to get a spare engine back to Plymouth. It is super motive power for the 13.45 from Penzance as 50 033+50 046 bring the train over Keyham Viaduct on the approach to Dockyard Halt, the background filled by the houses built mainly for Dockyard workers around the turn of the previous century. It is not such a clear view today. 50 033 had born its name since 26 June 1978, more of the much loved Ajax later on. 23 February 1982

50 034 Furious

The power of the 50s is well and truly on show in this low level study of 50 034 in the original plain blue livery powering through Dawlish over Colonade Viaduct – the shortest on the GWR - 1B83 1530 London Paddington - Penzance. Not a trespass shot either, a kink in the lineside fence and I could lean over a bit more then. 30 April 1977

Still in its original livery, but now bearing the name Furious it had received on 6 April 1978. Not the normal view at Exeter St David's as 50 034 is parked up in the yard after rendering assistance to a failed HST, 43018 being the Power Car behind the engine. I came across this by pure chance; the failed HST was on 1A29 1425 Plymouth-Paddington, beyond that I have no further details. 17 April 1982

50 035 Ark Royal

The finesse of the 50's meets the greatness of the Great Western. Brunel's highly embellished western portal of Box Tunnel viewed from the classic location of the A4 road bridge despite the never ending traffic. 50 035 emerges with the 1B29 1230 London Paddington-Penzance, diverted via Bristol with rural Wiltshire at its best for a backdrop. Need I say more?

Well perhaps a little more to draw the reader's attention to the unusual looking marker lights in the headcode box. These were a trial to see if a high intensity light might be more effective at drawing crossing users and track workers attention to an approaching train. In the end a single central headlight was settled on and fitted during the class's refurbishment. Ark Royal's headcode panel was returned to normal well before her trip to Doncaster for overhaul.
15 August 1976

Ark Royal, the first of the class to be named, and I had the pleasure of witnessing the ceremony at Plymouth Station on 17 January 1978. It is high tide at Shaldon, to the west of Teignmouth, as 50 035 proudly bearing the ship's crest above her nameplate passes by, working 1M22 1138 Plymouth-Manchester Piccadilly. It is nice to report that this view has only changed by the addition of a fence on the riverside of the line, easily overcome for photography by standing a few yards further to the right. 50 035 was the first of the class to be preserved, now in the care of the Class 50 Alliance, who are supported by The Fifty Fund, the publisher of this book.
02 April 1983

50 036 Victorious

It is a sultry summer's evening typical of the summer of '76 as 50 036, its head code box thinking it is 5036 Lyonshall Castle, heads through Dawlish Warren station working 1B73 1430 London Paddington-Paignton. The upside building to the left at this time housed the Dawlish Warren Railway Museum and Model Railway (1974-1984), an added attraction to the scene. Sadly this view point is no more as the station footbridge was removed a couple of years or so later, and the view would not just be the same now with the signal box and its semaphores all gone. 25 August 1976

We catch up with 50 036 a decade later on a similar working. Named on 16 May 1978, and now in large logo livery, 50 036 is working 1C44 1245 London Paddington-Paignton as it heads under a much missed landmark for many a traveller either by road or rail, namely the impressive signal gantry on the down line at Aller Junction. The latter would be gone within a year, and the view point itself would follow when the bridge was blown up on Christmas Day 2013 as part of the construction of the A380 South Devon Highway. The scene here is now totally unrecognisable and impossible to replicate. 20 September 1986

50 037 Illustrious

An interlude of unusual track for the class starting with 50 037 named Illustrious on 8 June 1978 on a working rarely photographed, and I only came across it by chance. 50 037 is heading along the East Coast Main Line at North Cowton about 8 miles north of Northallerton with the Doncaster Works Test Train comprised of a rake of Mark I coaches no longer in passenger service. This was a standard run to Newcastle and back for locomotives freshly out of Doncaster Works. 02 March 1979

A short but rare piece of line for the track bashers and the photographers as 50 037 cautiously creeps around the Westbury East Chord line of 1942 from Heywood Road Junction to Hawkridge Junction with the diverted 1420 London Paddington-Weston-Super-Mare. In the foreground are the rails of the route out of Westbury station in the Trowbridge direction which 50 037 will shortly join here at Hawkridge Junction. All Paddington-Bristol TM services this particular weekend were diverted via Newbury and the Hawkridge Curve to Bath (Bathampton Junction), but very few of them were locomotive hauled being mostly HST's, a class 50 slightly off its normal beaten track was the highlight of the day. 18 April 1981

50 038 Formidable

We have seen 50 035 emerging from the west end of Box Tunnel with its grand facade. It is another story at the eastern end of the tunnel; London direction trains emerge from the darkness through a portal no different to any other tunnel mouth. This one is hidden in a fairly deep cutting with no convenient road bridge or public access for Brunel to show off his handiwork. Having emerged from Box Tunnel 50 038 working 1A56 1130 Bristol Temple Meads-London Paddington is captured on the approach to Corsham from the road bridge between Valley Road and Potley Lane. 15 August 1976

I do not have many pictures of the Cotswold and Hereford route, a line I have really not had much cause to visit for some reason or another, hence my inclusion of a dull weather picture to broaden the illustration of the range of services worked by the class in their halcyon Western Region days. 50 038 was named on 5 May 1978, and is seen here taking the gentle curve at Yarkill near Stoke Edith with 1A25 1615 Hereford-London Paddington. 12 June 1983

50 039 Implacable

Taking the returning holiday makers home to face the realities of the commute, looking resplendent in the original pale blue livery, 50 039 on 1A55 1000 Paignton-London Paddington makes a fine sight accelerating along the Berks & Hants route at Lavington in rural Wiltshire. Whenever I travel this way, there is always a land mark to look out for at Lavington where the railway crosses the A360 road, and that is the curiously named pub The Chocolate Poodle. I do recall taking its refreshments on this hot day, and must find an excuse to go back some time. 24 July 1976

From the beer of the Chocolate Poodle to the Cider of Somerset as 50 039 (named 20 June 1978) looking fine in the large logo livery with a neat formation of Mark 1 stock hurries past the Taunton Cider Company at Norton Fitzwarren with 1M78 1127 Paignton-Liverpool Lime Street. A summer Saturday duty taking the holidaymakers home, this time those bound for the reality of life in the North West. These trains were all part of the magic of the railways in those days. 11 June 1983

50 040 Leviathan / Centurion

50 040 named Leviathan only two months previous on 15 September 1978 observes the St Austell stop for 1A13 12.20 Penzance-Paddington. The view is from the Carlyon Road over bridge. To the right the then by this time rare scene of a station goods yard still in use, coal traffic was handled here until 1985. From 1970 to 1984 the station yard in the summer season also doubled as the Motorail terminal, the coal merchant had to be on the ball to clear his delivery before the holidaymakers and their cars took over. Now it is an enlarged car park. The scene here looks much similar today, the footbridge was replaced by a more modern structure towards the end of 2014, the original being despatched to the Helston Railway for preservation. 13 November 1978

It is now 21 August 1982 and 50 040 has been refurbished and painted into large logo livery, seen here passing through Dockyard Halt, the rear of the train still on Keyham Viaduct, with 1M83 1050 Penzance-Manchester Piccadilly. This engine would become the second of the class to be renamed. The last HMS Leviathan had been decommissioned in 1920, while eight ships and a number of shore stations had carried the name HMS Centurion. In 1987 when 50 011 was withdrawn, HMS Centurion was the name of the Royal Navy pay centre at Gosport which had adopted 50 011. Consequently to maintain the current naval connection, BR transferred the Centurion name to 50 040.

50 041 Bulwark

Another member of the class that rarely appeared in my lens in pre Network Southeast days is 50 041, seen here passing the once attractive location of Berry Farm between St Germans and Bethany with 1V71 0734 Lees-Penzance "Cornishman". This was a favourite and easy to reach place for an after the early turn run out, nice and clear with a splendid backdrop of the Southeast Cornwall countryside. Sad to say the view is now spoilt by the trees that now exist here, and the head on view hampered by a colour light signal installed when the signalling was upgraded in the area in April 2018. 23 March 1982

50 041 was named on 8 May 1978 and looking a lot cleaner and tidier than the above picture of 15 months previous passing Narroways Hill Junction to the north of Bristol with 1S39 0735 Plymouth-Glasgow Central/Edinburgh Waverley. This is the start of the climb to Filton Junction on the quadruple track section, reduced to double track the following year and restored back to four tracks in the autumn of 2018. To the right is the still extant branch to Avonmouth and Severn Beach curving away from the main line. The gasometer (left), that was adopted by the fans of Bristol Rovers who's football stadium it dominated, is, like their Eastville ground, long gone. The footbridge is now cadged and a mobile phone mast stands in the way making a present day view like this close to impossible. 02 June 1983

50 042 Triumph

A flashback to the early days of the class on the Western Region as a very smart 50 042, in the original pale blue livery, descends the western flank of Dainton Bank at Newhouse Barton with 1B65 1330 Paddington-Penzance. This was one of the first trains on the West of England route to be routinely turned over to class 50 operation. 20 July 1974

The engine gained its name on 4 October 1978, and again looking very smart this time in the large logo livery, 50 042 is seen here in charge of the lengthy 1V49 0631 Dundee/0750 Glasgow Central-Penzance. The view captured on the approach to Gwinear Road at Sandy Lane, the lush surroundings of this part of West Cornwall looking at their best on this lovely summer evening. My local spy tells me the replacement concrete bridge that exists here now is a high one, and that lovely view is spoilt -like so many more- due to growth. Happily the engine survives in preservation and in Cornwall being based on the Bodmin & Wenford Railway. 04 June 1988

50 043 Eagle

Early days on the Western Region as a smart looking 50 043 ascends Dainton Bank, hard at work on the 1 in 37 approach to Dainton Tunnel passing the Stoneycombe Up distant signal with 1V91 1320 Liverpool Lime Street-Plymouth. It was unusual to get lovely evening light on a Bank Holiday Monday!! One regular problem with the class that plagued them throughout their lives was flashovers. On good authority from the then top management at Laira, I am told 50 043 was a particularly troublesome engine in this respect often giving cause for concern. 26 August 1974

In contrast to the express duties they were built for, a more humble outing for 50 043 named on 28 June 1978. Five years later to the month, as the HST's eroded the prestige express workings 50 043 waits to depart from Platform 5 at Plymouth with the 1557 stopping service to Penzance formed of 4 Mark I coaches, a delight for the enthusiast to enjoy a leisurely ride through Cornwall. A far cry from racing over the Northern Fells.

50 044 Exeter

This is 50 044 before it became Exeter in Exeter at Exeter West. A perfect pass as to the left 50 044 heading 1B65 1330 Paddington-Penzance greets 46 008 slowing for its Exeter stop with 1C04 1232 Penzance-Cardiff. The trains are crossing the relief flood channel built after the extensive flooding by the river Exe in these parts in 1960, The rear of 50 044's train would be on the adjacent bridge over the river itself. The effects of the drought are clearly evident; it would be a different story when the rains came. A couple of landmarks stand out. Above 50 044 St David's Tunnel on the sharp climb from Exeter St David's to Exeter Central is visible. Behind 46 008 towering above the houses is the spire of St David's Church. As a matter of interest, in the three hours spent around Exeter West that day in the era when summer Saturdays were far more interesting than now, and of course all the long distance trains were locomotive hauled, I saw 32 trains with seven different classes of motive power. 03 July 1976

Moorswater is the most graceful of all the Cornish Viaducts, the stone structure brought into use 25 February 1881 to replace the former timber viaduct whose masonry piers still stand on the south side. This angle is only available in high summer when the sun comes around far enough to light up the northern side of the structure. 50 044 had received its Exeter name on 26 April 1978, here we see it a couple of months later heading west over the 147 feet high Viaduct, well above the valley of the East Looe river with 2B36 1745 Plymouth-Penzance. I do recall not long after the engine had been given its name, one senior member of Exeter City Council proclaimed in the local press his pride in the city being the only one to have an engine bearing its name plying the rails of the south west. It was much to his and the chagrin of the Council it was pointed out the engine was actually named in honour of the Warship famed for its exploits in the Second War, the Battle of the River Plate in particular, rather than the city from which the ship had taken its name. 17 June 1978

The rhododendrons are in full bloom as 50 045 looking smart in the plain blue livery with its nameplate acquired 12 April 1978 is captured a couple of months after this event passing Treesmill on the descent to Par from Treverrin Tunnel. Achilles is bringing more holiday makers down to enjoy the delights of Cornwall with the 1V76 0845 Liverpool Lime Street-Penzance. This view can still be enjoyed today, but not sadly very often with a 50. 03 June 1978

A train off its normal Beaten Track: 50 045 looking fine in large logo livery with NSE stock passing St Denys with 1V17 1640 Waterloo-Exeter St David's, all services on the route on this day diverted from Worting Junction to Salisbury via Southampton due to engineering work. Look behind the second to fourth coach and the route to Portsmouth via Netley can be seen diverging to the right. 02 July 1989

50 045 Achilles to the Rescue

Now many will ask why on earth is a picture of a classic scene of GWR 4-6-0's 5051 Drysllwyn Castle+4930 Hagley Hall climbing the last lap to Dainton Tunnel in a book dedicated to the class 50 Diesels. Well, in every class of locomotive there is always one with a unique story to tell, such is the case for 50 045 when routine duty was sidelined for that unexpected call to arms leading to a place in railway folklore.

5051+4930 were working 1Z47 0902 Paddington-Plymouth Great Western Limited on 7 July 1985. Leaving Newton Abbot at 1750 it was running 50 minutes late. More woe was to come as the train stalled on Dainton Bank and efforts to restart it were in vain. This led to serious service disruption on the down main line. 50 045 which had been working 1C49 1445 Paddington-Plymouth was detached from its train at Newton Abbot and despatched to assist the failed special service. Thus the very rare – and I stand to be corrected on this one – and only time a class 50 was sent to the rescue of a steam special. Here we see 50 045 on the rear of the train at 1927 (now 2¼ hours late) pushing the ensemble up to the summit where the 50 was detached and the steam made its way down to Totnes. 50 007 was worked light engine up from Plymouth to return the special to London. 50 045 went back to Newton Abbot to resume its path on 1C49 having played its part in a piece of history. This had been my first encounter with a class 50 and a steam locomotive since the summer of 1968.

50 046 Ajax

The mighty Ajax: an engine very popular with the class 50 fraternity, regarded by many as one of the best performers of the whole class. It is a pity the locomotive did not join the ranks to survive in preservation, at least one of its cabs came to reside on the Plym Valley Railway for a while. Looking very smart in the plain blue livery with its nameplate (applied on 11 October 1978), Ajax is in full flight as it powers around the curve at Shaldon heading for Newton Abbot with 1B84 1327 Paddington-Paignton. The tide in the River Teign is low, stranding the boats moored here. 20 June 1981

How we like to remember Ajax, looking smart in the large logo livery in full flight through our green and pleasant land. A high level view of Combe St Stephen Viaduct between Burngullow and Grampound Road as 50 046 makes light work of the six coach 1M45 1202 Penzance-Liverpool Lime Street. In November 1991 Ajax was sidelined as it was due a D Exam for which NSE wouldn't authorise the expenditure. Following agreement to reinstate celebrity locomotive 50 007 for railtour duties from NSE's West of England Route Director, Peter Field, the power unit and bogies from the now withdrawn 50 046 were exchanged with 50 007 bringing an end to this popular machine. 01 October 1988

A reminder of the early days of the class in Cornwall; there is a fine display of gorse on show at Jollys Bottom to the east of Chacewater as plain blue 50 047 comes over the ridge powering 1M22 0915 Penzance-Manchester Piccadilly. An eagle eye will detect at the rear of the train to the left, just in front of the occupation bridge is the colour light Chacewater down distant signal, removed in 12th June 1977 when the box closed. 08 May 1976

It is very early in the morning – the only time of day this angle can be caught with the sun on the north east side of the Royal Albert Bridge from the Devon side. It is 0650 as 50 047 named Swiftsure named 26 May 1978 brings the penultimate locomotive hauled prestige 0505 Penzance-Paddington "Golden Hind" over the River Tamar on a gorgeous Thursday morning The service would the following Monday be turned over to HST operation. Behind the engine on the Cornish side of the River Tamar, Coombe-By-Saltash Viaduct glows in the early morning light; by coincidence our next port of call. If anyone is wondering what I was doing running around at this early hour, it was on my way home from the night shift. 02 August 1979

50 048 Dauntless

Passports have been checked so over the border to the other side of the River Tamar for a train that only ran on Thursdays, and only photographable in these parts late on a summer's evening. Looking in need of a clean 50 048 heads the heavily loaded 6042 1840 Burngullow-Sittingbourne, conveying clay slurry for Bowaters, over Coombe-By-Saltash Viaduct in the last of the evening light. Assistance from Plymouth over the South Devon banks would have been provided. The return empties came back on the Monday night running through this part of the world in the small hours, rarely seen and photographed. 19 August 1978

50 048 (named 26 May 1978) heads away from Lostwithiel, passing the creamery behind the tree to the right with 1M01 1610 St Austell-Kensington Olympia Motorail. There is an impressive line up of milk tanks in the creamery siding, but do not get misled. The following week the milk traffic from the West of England to London ceased. This was brought about, not by any fault of BR, but due to a combination of changing demand, and the capacity of Home Counties farmers to supply the capital's need for milk. The tanks were gathered here for storage in case of any policy reversal, which other than a short lived Swindon-Carlisle service never came to reality. 04 August 1979

50 049 Defiance

A classic Western Region early days of the class 50s scene where the signalling and infrastructure had not really changed since the days of steam, and the goods yard was still busy handling the traffic it was intended for. 50 049 captured approaching Newbury Middle Signal Box with 1B93 1630 Paddington-Paignton. A much changed scene here now with the area controlled from the Thames Valley Signalling Centre at Didcot, no goods yard and the view ruined by the electric wires. How can such a timeless scene be no more? One small change to the scene would occur on 2 May 1978 when 50 049 was named. 24 July 1976

All railway photographers have their own bogey engine, one that is very camera shy. After seeing it at Newbury in the picture above I saw '49 again just a couple of months later. I then had to wait until it became 50 149 before I caught up with it again. Thinking it is now GWR Castle 5049 Earl of Plymouth it is seen passing through the former Grampound Road station with 1M39 0935 Penzance-Wolverhampton. Grampound Road station was 2½ miles from the village it purported to serve and was closed on 5 October 1964. A community had grown up around the station over the years; this has in more recent times seen much expansion to the point there is a case for reopening the station. Note my little 100E car parked up outside the road entrance to the former goods yard. 25 September 1976

50 149 Defiance

As the frontline passenger train workings for the class decreased, a decision was made to evaluate the 50s for freight duties. In August 1987 50 049 was at Laira Depot repainted into Trainload grey livery and equipped with regeared traction motors and axels to provide higher torque. The engine became the first member of the class to be renumbered for a second time when it become 50 149. But despite the new number the Defiance name was retained.

The experiment was not a resounding success and within two years the engine reverted to its former identity resuming front line passenger work. Whilst 50 149 did not come up to expectations for example on trials with the Mendip stone trains, once back in the west country the engine worked locally from St Blazey Depot serving the needs of the china clay industry to which it was best suited. Seen from the A38 Plympton by Pass, 50 149 with the 1140 Tavistock Junction-St Blazey goods at the start of its journey. The newly introduced CDA wagons for the clay traffic look very smart at the rear of the formation; this traffic of loaded china clay from Marsh Mills to Fowey Docks would run for the last time on 29 April 2008. As a matter of interest, the CDA wagons would remain in use on the internal Cornish china clay circuit until the last ran on 11 August 2023. A further sign of the changing times is the 4 coach rake used for the local workings withdrawn due to the impending Sprinter takeover of these services. Tavistock Junction Yard was still a busy place at this time as can be seen with it full of freight traffic, most of which would vanish within the next four years and now what is left as an engineering depot is a mere shadow of former glories. 22 April 1988

While it was a freight engine, 50 149 saw minimal use on passenger duties. The most remarkable of these being its appearance on an Inspection Special which brought the unlikely sight of a class 50 on the Gunnislake Branch, a line where the number of appearances by the class can be counted on the fingers of one hand. 50 149 pauses at St Budeaux Victoria Road with Saloon TDB999506 in what was once the up platform of a double track main line to Waterloo with 2Z70 1100 Plymouth-Ernesetle Inspection Special.
08 November 1988.

50 049 Defiance (Again)

Defiance would become the only member of the class in the BR era to be renumbered for a third time. After a couple of years of being in the freight sector as 50 149, the engine was restored for passenger use resorting back to being 50 049 and came under the watchful eye of NSE. In the early morning light 50 049 emerges from the 265 yard long Kennaway Tunnel under Lea Mount slowing for the Dawlish stop with 2C08 0712 Newton Abbot-Exeter St David's. The train ran as empty stock from Laira Depot to form this fill in turn taking the early morning commuters to work in some style (and comfort) before forming 1034 0811 Exeter St David's-London Waterloo, taken on by 50 046 on this occasion. With the construction of the new sea wall here opened in May 2023, this precise scene has seen some changes for the better. The commuters now take their luck if they get a class 150 Sprinter or a class 165/66 Turbo, the latter exiled from the Thames Valley. For those of you aware of the time of day, yes I did get up very early to be in position for this working. Had to be done as the class 47s were now encroaching on these duties and my friends at Laira Depot the night before had assured me my very early morning drive would not be in vain. 25 May 1991

After ignoring my camera lens for many a year, 50 049 redeemed itself when it became part of the NSE fleet. Here it is pulling out of the depot at Laira Junction with the 1540 empty stock to Exeter. It was rare to see a move of stock from the site of the old steam shed as opposed to the carriage sidings to the left. The economics of having the depot for the Waterloo-Exeter route 50 miles to the west, and the high costs of empty stock moves, surely helped make the case for new multiple units and a depot at Salisbury mid way on the route. 05 May 1991

50 050 Fearless

A historic first as the pioneer member of the class becomes the first to work a passenger train on the Falmouth branch. The summer timetable of 1976 saw a new service introduced on summer Saturdays namely 1A79 0910 Falmouth - London Paddington, the stock working in on 2B70 0822 from Truro..The train would run for three seasons. In reality rather than for any commercial need, it was a useful way of saving space at Long Rock Depot, Penzance, during construction of the HST Shed. The stock came off the 1827 Fridays only Paddington-Truro, the engine off the 0300 Plymouth-Penzance vans then running light back to Truro to collect the coaches. 50 050 gleams in the morning sunshine at Falmouth waiting to depart with the inaugural 1A79 on 8 May 1976. It would be a further two years before the engine was named on 23 August 1978. The station would receive its "Docks" suffix on 15 May 1989 and the view now is impossible due to a block of flats standing where I stood to take the picture. 08 May 1976

Our pictorial journey is now heading to Newquay to start the final chapter. As we head from coast to coast soon to be an hourly Falmouth-Newquay Cornish Metro service now in development, a quick jump in the car to capture 50 050 working the inaugural 1A79 0910 Falmouth-Paddington crossing the nine arches of Carnon Viaduct . The Falmouth Branch former Brunel timber viaducts were the last in Cornwall to be replaced by masonry structures, the impressive one here coming into use as late as June 1933. Note the masonry piers of the original to the north side of the structure. 08 May 1976

Chapter 3: The Last Hurrah

"And the last will be first and the first will be last" to quote from scripture. When the class were renumbered in the spring of 1974 under the TOPS system with the class number forming the first two digits and after a space there followed 3 more digits, the system for some reason could not cope with three 0's for the first of any class. Thus D401 became 50 001 becoming the top of the table with the rest following in numerical sequence. Thus D400 had to go to the bottom of the pile and take the first available number 50 050 becoming the last in the numerical order. Double heading was very rare on the Newquay Branch. I was a passenger on this train 1B48 0948 Paddington-Newquay joining at Plymouth with 50 001 in command and in fine fettle. For some reason, I can only assume for testing purposes, 50 050 was attached at Par for the run over the branch, and it was the fastest run I ever had from Par to Newquay taking just 45 minutes. 50 050+50 001 seen on arrival at Newquay with the Gasometer once such a local landmark providing the back drop. Both are in the original livery but different shades of cleanliness. I wonder how often they had paired up in their days racing over the Northern Fells. They ran round the stock separately and 50 050 led back to Par on the 2B72 1658 departure where it was detached and made its way back to St Blazey shed. The next time I saw double headed 50s here was, as seen below, to mark the end of their reign on the line, and leads into the story of the decline and fall of the class. 03 September 1977

From the heyday of the class at Newquay to the end of booked locomotive working to the North Cornwall coast resort, a good time and place to commence that last hurrah, a phase which would see run down and withdrawal. In the latter years of the 1980's, the biggest surprise was the creation of Network Southeast in 1986 who took sponsorship of their fleet for working Thames Valley services and the Waterloo-Salisbury-Exeter route. The engines in reality remained in common WR use even as the front line work for the class was reducing. The HST's now bore the brunt of the London duties and by this time had made serious inroads into the cross country services. The last seven years of the class 50 would see many last workings and farewells as the HST stranglehold was completed, also that of the Sprinter revolution, modernisation of NSE services and by 1990 their reliability and costs of maintenance were giving cause for concern. A watershed occasion and place to mark the start of this final hurrah is here at Newquay where 50 035+50 034 both looking immaculate in NSE livery await departure with 1Z40 1740 Newquay-Paddington, a special run by BR to mark the cessation of locomotive hauled timetabled trains over the branch the previous day. In the pages that follow there will be more last days and workings as the scene changed, coverage of the NSE era, this livery became almost as common in Cornwall as it was say at Woking, Then there are the farewell rail tours – and there were plenty of them – taking the class back to their old haunts, off the beaten track and even to pastures new. These tours were popular and recalled by many as they were great occasions forging friendships and memories that last to this day. I could not cover them all but did a fair few of them. Enjoy the fun as the story is pictured and told in chronological order. 04 October 1987

A scene that would soon change as the summer timetable of May 1988 would see the end of the Class on secondary workings in the West of England as these were turned over to Sprinter operation. What was now a sorry replacement for the legendary 0800 Bristol-Penzance train with its buffet always behind the engine, 50 023 – the first to be repainted into NSE livery – hurries past Tavistock junction with 1C11 0645 Swindon-Penzance. One has to wonder how adequate was a four coach train was without any refreshment facilities covering such a distance serving commuters and day trippers amongst others along its journey. Perhaps it was to get the travelling public ready for what was to come in the immediate future with the two car sprinter on the horizon. A good day for me to be stood here though, as we have seen 50 149 followed on a freight for St Blazey. 22 April 1988

After the loss of the through summer Saturday trains to Newquay, the next milestone in the erosion of work for the class would come at the start of the summer 1988 timetable when as from Monday 16 May all of the local main line services in Cornwall and much else of those further east were turned over to Sprinter operation. On the last day of Cornish locomotive hauled locals, 50 039 and 50 042 double head 2C83 0911 Penzance - Plymouth captured here passing Shillingham Manor about half a mile to the west of Forder Viaduct. This service was often provided with super power as it was a convenient way to return a spare engine back to Laira off the overnight services into Cornwall. Back then there were the down sleeping cars as well as the seated overnight service from Paddington, the 0300 vans from Plymouth, the postal, and the newspapers. Now there this just the Night Riviera which just waits on Long Rock Depot until it comes back on Sunday evening and the authorities pray the 57XXX will start after its sojourn by Mounts Bay. This angle is now completely lost in the forest which now exists here. 14 May 1988

A climb up the hill for the view of the 77 feet maximum height Derrycombe Viaduct in the beautiful setting of the Glynn Valley (beware of the adders which was why many us did not venture up here that often) for 50 035 going well up the grade in charge of 2C84 1213 Penzance-Plymouth. 14 May 1988

I felt it was fitting to continue the theme of capturing these last local locomotive hauled trains on a classic Cornish Viaduct, so for the penultimate service a further chance to enjoy the most gracious of them all complete with the piers of the original Brunel timber structure on show. It is a mixed rake of colour and stock for 50 039 heading west over Moorswater Viaduct with 2C72 1520 Plymouth-Penzance. 14 May 1988

A little piece of history as 50 042 passes over Liskeard Viaduct (note the Looe Branch descending into the valley beneath the second and third coaches) with the final down locomotive hauled local service, the rather short three coaches only 2C74 1557 Plymouth-Penzance. More history was also unfolding at this moment in time, as the very nice aged couple who not only gave access to their garden, provided afternoon tea as on their portable television we then watched Wimbledon lift the FA Cup at Wembley beating Liverpool 1-0, the only time they would win the trophy. 14 May 1988

Double headed class 50s on freight workings were rare enough at the best of times, and almost unknown in the Network Southeast era. 50 041+50 044 are indeed exotic motive power for the 1505 St Blazey-Gloucester seen here at Tresulgan to the east of Menheniot. The short but interesting goods is well within their capabilities. I am not too sure NSE would have had such workings on their agenda when they sponsored the repainting of their fleet in to their own brand. 22 June 1988

It was only on summer weekends one saw much class 50 activity on the main line through Devon and Cornwall, so attention was turning to the Waterloo-Salisbury-Exeter route now under the control of Network Southeast. The cows graze on blissfully unaware of the drama to be seen on the nearby rails as 50 027 1O55 1607 Exeter St David's-London Waterloo passes by with Sherborne Castle standing proudly on the hill opposite. This is where Sir Walter Raleigh is reputed to have smoked his first pipe of tobacco. A tradition I have been proud to follow!! 24 July 1988

One of the few main line services in Cornwall still to be locomotive hauled as 1988 drew to a close was the direct train from Cornwall to its two parts of Bonnie Scotland; it would become an HST next time the timetable changed with no split portions, one would have to change for Glasgow and it would of course become half the size, with many complaints of overcrowding. I will leave it to your imagination the passenger comfort and comments when in 2002 Virgin Trains declared the 4 coach Voyager would be sufficient. This train was booked for class 47 haulage, but lady luck shone upon me this lovely autumn morning as I stood high on the hill looking down on Lostwithiel Golf Course as NSE provided the power for InterCity. 50 002 appeared on the lengthy 13 vehicle 1S71 0817 Penzance-Glasgow Central/ Dundee. Magnificent lengthy trains like this are now few and far between. 01 October 1988

On the same day as 50 002 passing Lostwithiel Golf Course on the previous page, another fast disappearing sight were the summer Saturday extras from Paddington. On the final day of the season, NSE in Mid Cornwall and almost a full matching set for 50 034 on 1C20 0902 Paddington-Penzance dropping down the 1 in 70 grade near Trebullan, just to the west of Burngullow. The main line here had been since 1986 a single line section from the latter to Probus, it was redoubled in the autumn of 2004 at great expense. A rural scene in the early tinges of autumn, the westerly limits of china clay country with the Blackpool Clay Works at Burngullow visible on the skyline. Obviously this is an unrepeatable shot, but still worth coming this way for a pasty at the nearby St Stephen's Bakery, one of the best in Cornwall. 01 October 1988

Off the beaten track for the first picture to take us into 1989 with the view from the Christchurch road bridge as 50 017 rounds the curve at Virginia Water from the Woking direction to join the route from Reading and Staines with 1051 1220 Exeter St David's-Waterloo. This train and indeed all others this particular weekend diverted due to engineering work. 11 March 1989

It is time I introduced in to the story what would be one of my favourite trains and one I would put in much effort to get in the frame. Briefly, NSE introduced in 1987 a weekdays 0940 Plymouth-Portsmouth Harbour with a corresponding 1159 Portsmouth Harbour-Plymouth utilising resources from the Waterloo-Exeter workings, this involving a locomotive hauled train each way over the Portsmouth Direct route for positional purposes. These would only last a couple of years but at my end of the line the 0940 in particular was ideal for morning photography, often with the added bonus of double heading as a way to get motive power up to Exeter. It was also perfectly positioned at Plymouth Station in Platform 5 at a convenient time for the early turn Supervisor in the ticket office at Plymouth to pay a visit to the station offices then at the east end of Platform 4. What a coincidence on this glorious early spring morning to find 50 048 fuming well on Platform 5 awaiting departure with 1O28 0940 Plymouth-Portsmouth Harbour. Looking at the date, perhaps Irish eyes were smiling on me!! 17 March 1989

Classic motive power and a mixture of coaching stock liveries for 50 048+50 044, 1O28 0940 Plymouth-Portsmouth Harbour at Crabtree crossing the River Plym at high-tide, the view from the A38 Plympton By Pass. With just five weeks of these workings remaining, considerable effort was put into recording them on film as will be noted the pages that now follow. 07 April 1989

A grand day out, my own personal rail tour – how I wish one could do this today. A very enjoyable ride on the 0940 throughout, the picture gives the impression of the train arriving at Portsmouth Harbour. It is actually the empty stock going out to Fratton to run round with 08 830 out of sight leading on the other end. 50 001 at a rare location for a picture of one of these I suspect. 11 April 1989.

My personal rail tour continues joining 1P40 1530 Portsmouth Harbour-London Waterloo, a run over the Portsmouth Direct route behind one of these was an absolute must before this rare working finished. 50 001 working 1P40 pauses to pick up passengers in the rather cramped and always busy modern looking high level platforms at Portsmouth & Southsea; the recently built station canopy of the early 1980's looking very smart in NSE colours. With my desire to get the picture at Portsmouth's main station, I had a word with the driver before leaving Portsmouth Harbour who agreed he would not leave the High Level until I was safely back in the front coach, note the open door.. He had told me to be quick as they were always in a hurry to despatch trains there, mission accomplished and another rare location in the bag. 11 April 1989

Classic Waterloo in the early evening sunshine as 50 049 simmers away prior to heading away with 1V17 1706 to Exeter St David's, despite the lack of people in the picture, be assured the train was full of commuters, the further west it went the emptier it became; a good end to a perfect day out. 11 April 1989

An angle now impossible for me as I do not have track side access now I am retired. 50 027 looking impressive passing under the private (sadly access to it is forbidden by the land owner) farm occupation bridge immediately to the east of Sparkwell Bridge, at the summit of Hemerdon Bank with 1028 0940 Plymouth-Portsmouth Harbour. If I were to stand in this position again, it would be different as the down loop and its signal P306 have vanished without trace, a modicum of growth and lineside clutter would spoil the view. 26 April 1989

As its forthcoming demise edged ever closer, I was taking every opportunity if the weather was decent to cover the Portsmouth in one form or another. This time it is the down working departing Totnes, the view from the Western bypass road bridge when it was a lot clearer than it is now, the sun obliging as the train left the station for this colourful view. 50 048 1V12 1159 Portsmouth Harbour-Plymouth 27 April 1989

Classic NSE in a corner of South Devon: A beautiful spring morning, two clean engines and fully matching stock for the impressive sight of 50 018+50 041 1028 0940 Plymouth-Portsmouth Harbour on the climb between emerging from the woods to the summit at Hemerdon. Well worth the effort to get up early (by my standards) and enjoy such a sight and sound. 28 April 1989

I had certainly chosen the right morning to position myself at Stoggy Lane, but I have to admit I knew this was coming, inside information was always very helpful in those pre-mobile phone and internet gen days. A mini cavalcade of class 50s; 50 021+50 042+50 029, on their way from Laira Depot to Exeter with a nice mix of colour of liveries. With recent track renewal and ballast dumping, this location had only been recently cleared allowing photography for the first time in many a year. Fortunately the tenant of the allotment was very friendly with a come anytime you like invitation, one accepted more than once as will soon be proved. 05 May 1989

For the last week of operation of the weekday Plymouth-Portsmouth and associated workings, I took annual leave, and accompanied by my much missed close friend Maurice Truscott, went to see the trains "at the other end" and get some photography on the Portsmouth Direct route. One downside to the trip was although we had free accommodation staying with Maurice's sister, her house in Gosport must have been the one the furthest from any pub in the town!! The other potential snag was being away for my forth wedding anniversary, a very understanding wife granting me the necessary "pass". We made it up from the West Country in time to capture 50 041 1V12 1159 Portsmouth Harbour - Plymouth rounding the curve at Redbridge to the west of Southampton taking the Salisbury route. 08 May 1989

The spectacular view from the A3025 Itchen Bridge as 50 029 leads 1O28 0940 Plymouth-Portsmouth Harbour around the curve on the approach to Woolston Station, a view which seems to have escaped many railway photographers. This is perhaps not surprising as in reality this was the only train of any interest along here especially at this time of day when the sun was in the correct place. 08 May 1989

Mission accomplished for day one of the trip and another rare location, 50 029 1P40 1530 Portsmouth Harbour-London Waterloo leaves England's only island city crossing waters of the Ports Creek approaching Portcreek Junction. In this land of multiple units be they diesel or electric, even at this period in time, locomotive hauled passenger trains here were few and far between. 08 May 1989

Activity at Fratton as 08 830 on the rear of this view has dragged 50 044 and its stock up from Portsmouth Harbour after working in on the 0922 from Waterloo. Both engines would then swop ends in a short period of intensive shunting around, which brings us nicely to the next picture. 09 May 1989

The view looking the other way and another classic Portsmouth area viewpoint at Fratton as 08 830, out of sight on the front of the train, leads the stock for the 1159 to Plymouth with 50 044 bringing up the rear down to Portsmouth Harbour for the return working. The running of these services no doubt caused some inconvenience and expense especially in what was now a unit dominated area, the provision of the 08 and its crew to twice daily pull the Plymouth train in and out of the Harbour was doubtless a factor in the economies of running the service. The matter that it served a popular and well used flow of traffic probably did not come in to the equation when the decision was made to withdraw these popular and very useful trains. 09 May 1989

Portsmouth at its very best and a superb location for railway photography rarely seen, 50 044 passing Canal Walk just after departure from Portsmouth and Southsea with 1V12 1159 Portsmouth Harbour-Plymouth. 09 May 1989

Another location which had to be covered as again this would be rare Class 50 photography territory was the Hamble Viaduct just to the east of Burseldon station. In this pleasant setting 50 041 hurries over the structure with 1O28 0940 Plymouth-Portsmouth Harbour. Our plans for the rest of the day then became subject to alteration. A fellow photographer we met here said he was now off to Lavant for the sand train with a class 73 which was running on this day, and he knew the people in the Quarry and access would be no problem. My apologies to the class 50 fraternity, as it turned out 73 002 on the sand train at Lavant is now just as rare and priceless as class 50s on the Portsmouth Direct line. Sometimes one has to make a difficult choice. 09 May 1989

After the unexpected and very successful diversion to see and photograph 73 002 at Lavant, took the long way home to Gosport to visit the classic spot at Totters Lane between Winchfield and Hook on the iconic quadruple dead straight track section of the LSWR main line with its two landmark over bridges in the background. Our reward was 50 023 in lovely evening light galloping past with 1L15 1734 Waterloo-Salisbury. 09 May 1989

Business on the Portsmouth Direct Line commenced the following morning in rural Hampshire with 50 018 near Rowlands Castle caught on camera at Comley Hill from the B2148 road bridge with 1P15 0922 London Waterloo-Portsmouth Harbour. A locomotive hauled passenger train passing through here was a rare sight at the best of times. 10 May 1989

Whilst 50 018 went in to Portsmouth and after its trip to Fratton to run round and then come out again, we made our way over to Burseldon for the southbound shot on the Hamble Viaduct by the marina. Sadly the weather was not so good on this day; the sun had deserted us and since as the train would only run for the next two days, a return visit was not on the cards. Despite the cloudy skies, 50 018 still made a fine sight heading west over Hamble Viaduct with 1V12 1159 Portsmouth Harbour-Plymouth 10 May 1989

A greedy extra, I was positioned not only for the viaduct shot but also for the broadside view going over the A27 road bridge, and was blessed with a very fortunate gap in the traffic. Just immediately beyond the bridge the left hand turn is the access road to Burseldon station which 50 018 is rapidly approaching with 1V12 1159 Portsmouth Harbour-Plymouth. Somebody may recognise themselves by the open window listening to the music from the carriage right behind the engine. This now a practice totally taboo, the hobby was more fun back in the days of common sense. 10 May 1989

Very rarely photographed is the wayside station that serves the nearby village of Hamble-le-Rice. In this rural setting 50 028 passes through Hamble Halt with 1O28 0940 Plymouth-Portsmouth Harbour. If the appearance of the station and in particular the waiting shed on the down, i.e. left, platform appears one of austerity, this is because it was a very late addition to the line being opened 18 January 1942, more to cater for those travelling to and from the recently opened aircraft factories than the local population, who had campaigned for years to be served by the passing trains. At this time the winds of change were on the horizon and there is a clue in the picture with the appearance of a third rail on each track. It would be a year and a month after my visit here that the line was electrified on the SR three rail system with electric services commencing in May 1990. In the process, NSE provided modern waiting shelters to replace those, seen here, dating from the Second War. A perfect example of how the railway scene can change in a short span of time. 10 May 1989

The panorama that is Guildford: Whilst 50 028 went in and out of Portsmouth, we made our way through the traffic to find another superb photographic location which probably does not get the attention it deserves as there are so few locomotive hauled trains to be seen around here. This is the view from Yorkie's Bridge, which takes a very convenient footpath over the railway, as 50 028 pulls away from Guildford with 1P40 1530 Portsmouth Harbour-London Waterloo. In the background amongst the plethora of Electric Multiple Units on display, 47 566 can be seen waiting to take the Reading line with a van train from Redhill, these like the class 50s here now also just a memory. 10 May 1989

By virtue of the geography, the shot of a train coming off the Portsmouth Direct route from Guildford joining the coast way route from Brighton at Havant would be an evening one, but since the only locomotive hauled train to perform this move came at about 1035, best to take advantage of a cloudy morning to avoid the awkward lighting conditions which would have prevailed. It is a very nice angle and another of those rarely seen on camera. 50 048 rolls around the gentle curve off the Portsmouth Direct line passing the Wessex Plant Hire premises to the left, and bringing the penultimate locomotive hauled 1P15 0922 London Waterloo-Portsmouth Harbour into Havant station. Worth getting your history books out for, as this is the site of the "Railway battle of Havant" where the London & South Western Railway and the London Brighton and South Coast literally lined up against each other, with men from both sides at one stage facing a standoff. 11 May 1989

This was not a good day for photography with dire lighting conditions so my apologies for not the best quality picture, but worthy of inclusion as it is so rare and the scene would be history by the time the film came back from the processors. Taken from the farm bridge just to the east of the station, 50 028 heads away from Rowlands Castle with the penultimate locomotive hauled 1P40 1530 Portsmouth Harbour-London Waterloo. Rowlands Castle station can be glimpsed behind the train. Then it was time to head for home after a very productive trip, and I did make in time to take the good lady down the village pub for that wedding anniversary drink(s). 11 May 1989

The final 1O28 0940 Plymouth-Portsmouth Harbour on a beautiful late spring morning, and the train certainly went out in style with both locomotives looking immaculate. 50 003 and 50 028 about to pass the 241 mile post at Stoggy Lane, on the lower reaches of Hemerdon Bank, provide the last motive power for what had become my favourite train. Adieu, thanks for the memory. 12 May 1989

The last 1V12 1159 Portsmouth Harbour-Plymouth, the down working did not go out in quite such a blaze of glory as the up one but 50 043 and a rake of matching liveried stock look the part passing over the Aish Emergency crossovers – which still exist, electronically released when required for use from the Plymouth Panel- just to the west of Brent. Here the railway runs just within the southern boundary of the National Park, NSE on Dartmoor!! To the top right of the picture note the landmark of Brent Hill. 12 May 1989

We have had classic Waterloo; this is classic Paddington as we remember it before the electric wires went up and the Elizabeth line came to be. 50 024 waits to depart with 1F52 1748 to Oxford ,to the left is the former Paddington Arrival Signal Box , a relic of the early 1930's resignalling scheme displaced in 1967 when the Old Oak Common Panel took over, then is use as the Paddington TOPS Office. Now it is no more. I have an affinity with Vanguard as I recall as a child circa 1954 going on board HMS Vanguard, the only Battleship not to fire in anger, when she was on display in Devonport Dockyard. 19 May 1989

Through wide open spaces that are now not so open, a panorama of the railway in the landscape. Running parallel to a free flowing M5 Motorway, 50 030 with 1A50 1020 Paignton-Paddington formed entirely of NSE stock running through the rural fields and countryside of East Devon. The view is from Upton Hill just to the west of Cullompton. The road bridge in the foreground is the A38 road, now devoid of most its traffic. The engine survives in preservation and this splendid view lives in the memory as it is now impossible due to growth. 24 June 1989

A reminder of the Paddington-Oxford Thames Valley services as 50 032 calls at what is now a very much changed scene at Reading with 1F57 1250 Oxford-London Paddington. To the left a 4VEP awaits at Platform 4A with the next service to Waterloo via Staines. Hard to believe at this time these were the only electrified rails to be found on the Western Region and the class 50s only had a couple more years of service on these workings. 02 July 1989

My moves that day were to try and capture Waterloo-Exeter services diverted via Southampton, as it turned out not entirely successfully. 50 049 passing Bevois Park on the approach to Mount Pleasant Crossing with the diverted 1O39 1548 Exeter St David's-London Waterloo. This splendid location between Southampton and St Denys is still a good place to visit for photography, and with the Freightliners amongst other things there is still a good variety of trains, but sadly not with one of these. 02 July 1989

One of my best pictures of a class 50 dragging a failed HST. A helping hand from NSE for Inter City as 50 037 assists 43006+43124 on the late running 1410 Paddington-Penzance out of Mutley Tunnel on the approach to Plymouth Station. I was the late turn Supervisor in the Plymouth Ticket Office this day, my meal break taken at a very convenient time. 14 July 1989

Network Southeast by the sea side and the Teignmouth sea wall at its very best: a colourful study of a smart 50 017 and matching stock about to enter Parson's Tunnel with 1086 0933 Plymouth-Brighton. There do not appear to be many people out enjoying the delights of this part of the coast on a glorious summer's day. I fear my legs would not now allow me to scale the rocks from the beach any more to reach this classic location. 15 July 1989

On the way home after a very successful day after wandering around Dorset in search of interesting trains, there was just time to call in at Honiton for the view from the footbridge in the evening sunshine as 50 048 rolls in with 1O43 1822 Exeter St David's-Waterloo, passing DMU P871 waiting with the 1835 Axminster to Exeter St Davids. 05 August 1989

For some reason there is a gap in class 50 photography of about eight months, partly due to trips to Scotland and other far flung places, highlighting also the fact this really was the final chapter of the last hurrah of the era as withdrawals and loss of work took their toll, and pictures of them became fewer. So it is fast forward to the following spring for an attempt at artistry with a spruced up 50 043 contrasting well in its smart NSE livery with a newly repainted welcoming genuine GWR platform seat on Platform 4 at Plymouth Station. 28 April 1990

As 1990 progressed, so did withdrawals, so began the railtours which marked the final four years of the class. Passing under the then decaying footbridge at Okehampton is 50 020 with the outward working of the Taw Tourer and Tor rail tour (1Z17 0445 from Manchester Piccadilly). Much to the dismay of many on board, the train did not actually live up to the Taw part of its name. It was running so late, the leg to Barnstaple was cancelled. The photographic party only found out in those pre-mobile phone days when waiting around at Lapford we realised it had lost its path on the single line. One of the waiting gallery made a phone call from a conveniently placed red phone box to control, where a friend was on duty. The updated position was relayed and there followed a dash through the lanes to Okehampton, we arrived at the station with literally 30 seconds to spare, the trials and tribulations of the railway photographer. 05 May 1990

What might have of been: 50 020 pauses for a photographic stop at Okehampton on its return from Meldon Quarry. From the stop blocks at Meldon there are just 21 miles of missing track between there and Bere Alston where the southern end of this former main line makes its way into Plymouth, now part of the Gunnislake Branch. Had those rails been left in place, the route via Okehampton would have been a valuable diversionary route for the vulnerable and of the troublesome sea wall section from Dawlish to Teignmouth, and there is no doubt the class 50s would have been seen here many a time. It was not to be and probably never will be. As far as I know this was the last passenger train working for 50 020 as it was withdrawn two weeks later and cut up at Boothe-Roe scrap yard, Rotherham in June 1992. Okehampton station was reopened to passengers under the Government's restoration of Railways Plan in November 2021 and now has a vey successful hourly service to and from Exeter. Before the service went hourly, the opportunity was taken to run a special out to Okehampton top and tailed by 50 007 (thinking it was 50 034) and 50 049 on 26 March 2022. The last class 50s to come here? Who knows? 05 May 1990

By the summer of 1990, class 50s on their way to Cornwall on service trains were no-longer a daily sight. I know not why this train ran, probably to clear a backlog of traffic for some problem that had occurred, 50 045 was turned out for a 1420 special Parcels to Penzance. Very convenient for the early turn Ticket Office Supervisor just after the shift had ended. The train is seen pulling out of the station at North Road West, the green grass denies any clue to the past here. The train is taking what was in its time the Cornwall Loop built in 1876 mainly for the LSWR to avoid Millbay – a long story. I am standing in what was the V of the junction for the Millbay line to the right where the Locomotive Service Point last used in 1963 existed. 16 June 1990

The Saturday Brighton had survived to run another year. An impressive sight leaving Plymouth on a glorious summer morning, 50 027+50 023 provide super motive power for 1086 0915 Plymouth-Brighton. By this time the class 47s were making inroads to the NSE services and the 50s were not now a daily sight in Plymouth. A pair in fine fettle was a treat. 14 July 1990

The Brighton again as the early autumn colours take hold and another rare angle taken from the remains of the former down platform at Plympton station closed 03 March 1959, demolished spring 1964. Passing the former Plympton Signal Box closed on 25 June 1967 and then used as a base for the Permanent Way staff, 50 012 is at full power just about to start the ascent of Hemerdon Bank with 1O86 0915 Plymouth - Brighton. There was not any public access to the former station at this time, one of my colleagues in the Permanent Way department had told me he would be there that morning, say no more. The former signal box was demolished on 19 November 1990. The shot is now totally impossible to heavy growth which also prevents any access in the first place. 15 September 1990

A dreary day weather wise in the Capital on my way to Bonnie Scotland but a picture which merits inclusion due to the rarity of a class 50 hauling such stock. From Canterbury House 50 026 is about to reach its destination with 1L14 1318 Salisbury-London Waterloo, the load 2 of 4TC sets. In simple terms, the latter were push pull stock introduced in 1967 as part of the electrification to Bournemouth being able to be hauled by suitably equipped 4-REP EMU'S and thence to Weymouth by class 33/1 diesels. After completion of the third rail all the way to Weymouth in 1988, some class 33's and TC stock found employment on services like Waterloo-Salisbury, and the odd one as far as Yeovil Junction, but very rarely were the class 50s seen on TC stock as this would involve a second engine to go on the back at Waterloo to come out again, and running round at the eventual destination. 17 October 1990

So to 1991, the last year the class would see squadron use. It began with 22 remaining in service and would finish with just 5, these being the celebrities kept primarily for railtours that would take the class back to some of their old haunts, pastures familiar and some not so familiar. These trips were enjoyed by many both on and off the trains, and the memories remain for us all. One small drawback for those on the line side was that the weather conditions were more often than not as favourable as one would have liked. As for main line duties, the Thames Valley services would succumb to the class 47s, whose presence on the Waterloo-Exeter route grew as the year wore on. It is a still night at Exeter St David's as 50 033 has arrived with 1V17 1655 from Waterloo. 20 January 1991

Things looked very much still as normal on this January night at Exeter St David's as 50 033 has run around its train and waits to take the empty stock to Laira. Having the western stabling and servicing point for the route, with all the associated stock and locomotives involving a trip of 50 miles each way did not do too much for the economics of the whole operation. I had come up to Exeter to ride back to Plymouth on the last Sunday 1645 Brighton-Plymouth, 50 046 doing the honours marking the end of another era, the first of the forthcoming changes. To the right Metro Cammell DMU P831 waits in Platform 1 with the 2110 to Exmouth, that in itself a scene soon to be as much a memory of the class 50s working the Waterloo trains. My note book tells me the exposure on 64ASA Ektachrome film was 30 seconds at F5.6. 20 January 1991

The first month of 1991 also saw the first of what would be many of the farewell to the class rail tours. I could not get out to cover them all but made a good effort mainly for those in Devon and Cornwall. It was disappointing the weather did not often recognise the importance of such events; the pictures are included for the nostalgic and some rare places value. It was a dull day for 1Z16 Cornish Centurian rail tour, my first reasonable interception of it was at Ringwell Bridge on the Falmouth branch with the 1238 Falmouth-Bugle leg of the trip. Top picture 50 015 heading into the cutting having just traversed Carnon Down Viaduct. Lower picture: 50 008 on the rear as the train heads away towards Truro. 26 January 1991

Still topping and tailing, the tour has now made its way on the Newquay Branch but due to pathing constraints going only as far as Bugle. A modicum of winter light as 50 008 leads the cavalcade on the causeway alongside the Par River at Rundles Walk, and below on the rear 50 015 glistens with the St Blazey down starting signal adding to the scene. Lots of heads leaning out on the carriage windows enjoying the thrash, one cannot enjoy that experience now, nor can the picture be taken either as the causeway is now one with a mini jungle. For the record, the train was 1Z16 2320 from Manchester Piccadilly arriving back almost 23 hours later. Such endurance tests were part of the fun. 26 January 1991

50 015 out of sight on Molinnis level crossing with 50 008 bringing up the rear departing from Bugle with the return Cornish Centurian Tour. This and the repeat tour of 4th May 1991, are the only known passenger trains booked to terminate and then start at Bugle, unless anyone knows better. This scene is now unrecognisable. 26 January 1991

Spring has sprung and this was a good day to be out the line with a camera as the class 50s outnumbered the class 47s, a situation that was changing rapidly. Classic location, classic engine as 50 007 in its fourth variation of green livery (a saga best left to Geoff Hudson to tell the story) is going well up Honiton Bank at Wilmington with 1V11 1202 Basingstoke-Exeter St David's. All Exeter-Waterloo and return services were start/terminate at Basingstoke this day due to engineering work. 30 March 1991

Seaton Junction was one of those places where if the sun was out, a decent picture could be obtained at any time of the day. It is in the late afternoon spring sunshine 50 037 is captured working 1V13 1402 Basingstoke-Exeter where there is evidence of past glories. The station closed 07 March 1966, the four tracks through it reduced to just one when the route was singled in May 1967. One reminder of the past still open here at this time was the Shute Arms Hotel, literally right outside the station opposite the former up platform. This once excellent hostelry closed in 2007 and is now a block of flats. A sad loss but it was not exactly near any centre of population. Even this angle is no more as the station footbridge has since been demolished. 30 March 1991

Looking the other way from the footbridge which survives as part of a footpath at the former Seaton Junction station there was wide open space, and until the late 1960's this was a hive of activity with quadruple tracks through the station, a branch line junction in the foreground and behind the train, the milk sidings, 50 018 makes its way back east with 1O38 1422 Exeter St David's- Basingstoke. 30 March 1991

Back to more rail tour action in Cornwall with Cornish Centurian 2. An impressive view of 50 008+50 015 from the top of Higher Town Tunnel with the rear of this lengthy train still coming around the curve from the main line to meet the Falmouth Branch at Perwithers Junction.

The train was booked to traverse the branch, but, sadly due problems with 50 008 that caused significant delays to the train, that was abandoned, so this is all the participants on this tour saw of the Falmouth Branch.

You will see it is the Falmouth Branch that carries straight on, as the Cornwall Railway was constructed from Plymouth to Falmouth.

The history here is quite complex, briefly: first here was the standard gauge West Cornwall Railway opening from its station at Higher Town to Penzance in 1852. The broad gauge Cornwall Railway arrived in Truro in 1859, extending to Falmouth in 1863. Mixed gauge was then laid on the West Cornwall and the first through train from Penzance to Paddington ran on 1 March 1867.

The Signal Box here was closed on 7 November 1971 and the Junction was rationalised, with the Falmouth Branch only accessed from the down main line which became bi-directional from here to Truro. 04 May 1991

The spring blossom is out in the hazy sunlight at Truro Viaduct, the view from a convenient car park. 50 008 has been placed to the rear of the train for the run up to Burngullow due to the requirement to top and tail on the Parkandillack and Newquay branches. 04 May 1991

Look at all those happy smiling faces leaning out of the train enjoying the fun and the noise – now of course against the rules – as 50 015 heads the Cornish Centurian 2 rail tour, a second version of the 2320 Manchester Piccadilly-Bugle which had run on 26 January 1991, off the Parkandillack Branch at Burngullow, this in the era of the single line from Burngullow to Probus. Compare with the picture of 50 006 here in Chapter 2. 04 May 1991

The tour did recover its schedule to run up the Newquay Branch as booked to Bugle and back (the second ever train booked to terminate and restart at Bugle?) but only with the class 50s at one end, this due to 50 008 being declared unfit to head the special on its own. 37 673 came to the rescue and is out of sight on the front as the cavalcade passes Luxulyan, and lady luck showed her hand as the sun appeared for just the very minute 50 008+50 015 on the rear came in to the lens. As a matter of interest, the authorities had chosen 50 015 to be painted into the Dutch engineering livery because it had snowploughs fitted, 50 008 into "Laira Blue" to represent it being on the books as a departmental engine. They always looked a good combination when seen together, nice for the photographers to win one now and again, it is usually the reverse situation. 04 May 1991

It is high summer in the last one of class 50 operation on the Waterloo-Exeter and associated trains, and low tide at Rockstone as 50 027+50 031 1O86 0925 Plymouth-London Waterloo (the apology for the Saturday Brighton now no more) pass by under the red sandstone cliffs. Locomotive hauled trains along the sea wall were by now in the minority, and class 50s getting few and far between, as the sprinters and HST's dominated most of the workings. Towards the rear of the train is the section of the Dawlish sea wall that collapsed in the violent storm of February 2014. It all looks so idyllic on a perfect summer morning, 25 May 1991

Much of the front line work for the class had ceased by this time, they found useful employment on duties such as stock moves. Coming home after a Diesel Gala on the Paignton and Dartmouth Railway, 50 008+50 015+D7017+D1010 are an impressive cavalcade passing Aller Junction running as 1050 Paignton-Bristol Bath Road Depot. There will no doubt by many a debate as to which is one's favourite type of engine!! Plenty of choice in this unique main line scene. 26 June 1991

The going away shot of the Paignton to Bristol Bath Road Depot cavalcade is indeed a sight for sore eyes, enhanced by that lovely just after a shower of rain lighting. Worth getting wet for. The viewpoint is no more, eradicated by the A380 South Devon Highway opened in December 2015. It is sad to note that the actual destination of this working, Bristol Bath Road Depot adjacent to Bristol Temple Meads station, the latter providing a grandstand view for generations of railway enthusiasts of the comings and goings to and from the shed, was closed 28 September 1995. The coming of the HST and Sprinter eras had considerably reduced the need for a separate locomotive depot, GWR concentrating its activities in one place at the nearby St Phillips Marsh Depot. The Depot was subsequently demolished and the site earmarked for redevelopment. As yet nothing has happened and the site is best described as an eyesore. How the railway scene has changed over the years. 26 June 1991

As the summer of 1991 progressed, the number of class 50s working on the Waterloo-Exeter route was well in decline and one had more chance of seeing a Class 47 than a Class 50. A few still soldiered on so it was still worth the effort. Returned to its original number to mark its celebrity status and popularity with the ever growing devotees of the class, D400 (50 050) was captured at Tallaton between Feniton and Whimple in the lush East Devon countryside, working 2V07 the 0745 Basingstoke-Exeter St Davids. 21 July 1991

Such was the infiltration of class 47s on to the Waterloo - Exeter route, this was my second attempt to capture a class 50 at the attractive location at Whitford just to the east of Seaton Junction. Here after following the valley of the River Axe for 13 miles, this is the closest the Salisbury-Exeter line gets to the sea about 3 miles distant before heading inland towards attacking Honiton Bank. 50 033 passing Whitford 1V10 1114 Southampton-Plymouth, and yes the growth now spoils the view. 07 September 1991

The matching NSE locomotive and stock when nice and clean looked very smart, and were not only a pleasure to photograph but also to travel on, the Mark 1A stock was very comfortable and one could open the windows and especially in the leading coach listen to the music coming from the head of the train. By this time in the Plymouth area, weekends were the best chance to capture them as they ran later in the day especially on a Sunday as opposed to the much earlier starts required for the weekday service. It is a glorious Sunday morning for Glorious and her matching train 1037 1105 Plymouth-London Waterloo for my on the way to work shot approaching the former Plympton station. One would never know to the left the recently built housing estate was once the station goods yard. 15 September 1991

Another slice of history: a truncated version of the through Brighton-Plymouth service had run on summer Saturdays for the 1991 season from Southampton. This ran for the last time on the final weekend of the summer timetable with 50 046 working the train into Exeter St David's, 50 017 attached at the other end for the leg down to Plymouth. I rode on this train from Exeter St David's to Dawlish Warren where the service was booked a stop not far short of 10 minutes for a down express to overtake. It was quite a dash from the platform to the over bridge for this historic shot of 50 017 coming out of Dawlish Warren with the last 1V10 1114 Southampton-Plymouth. Fortunately the train crew were a couple of Plymouth based good friends of mine, who agreed not to leave the platform until the out of breath photographic party gave them the necessary wave from the over bridge. In this day and age of delay minutes and close monitoring, somehow I don't think we would get away with a caper like this again. Hence I rode on and photographed the train; the end of locomotive hauled trains from the south coast to the south west had been recorded for posterity.
28 September 1991

Now very much in to the autumn of the class, their front line work had greatly decreased. They were still finding employment on special and one off duties, such as this mini cavalcade: 50 008+47 816+D7018 pass Harford (Palace Lane) just to the east of Ivybridge 6X58 1130 Laira-Taunton. 18 September 1991

More rail tour action, this yet another overnight endurance test from Manchester Piccadilly (departed 0030) behind a class 47, the 50s taking over at Bristol to firstly go down to Paignton. I have always wondered why so many would wish to leave the North West at such hours, but such was the dedication, enthusiasm, camaraderie and decent rolling stock for the hallowed motive power, a night on the cushions was a small price to pay, and I did a few in my time. Due to work commitments I could not deal with the morning's adventures around Torbay, I caught up with the special squealing on the curves on the approach to Luxulyan in a lucky burst of light 50 033 leading The Valiant Thunderer tour, 1Z37 0030 Manchester Piccadilly-Newquay via Paignton. Just look at all those heads leaning out enjoying the fun; this is what it was all about. Arrival back in Manchester would be a shade just under 24 hours after the tour had set off; I wonder how many were still awake? 23 November 1991

The chase was on, the going away shot with 50 033 out of sight leading the train and 50 015+50 008 bringing up the rear as the ensemble pulls away from St Dennis Junction on the last lap to reach the Atlantic Coast at Newquay, the viewpoint what was then the A30 road bridge. 23 November 1991

50 033 on the stop blocks at Newquay, a shadow of itself reduced to just one line serving what was Platform 2 in October 1987. Compare this to where we saw 50 050+50 001 in Chapter 2. At least the station still has Cornwall's longest platform. 23 November 1991

Sun set on the Newquay Branch: 50 008 + 50 015 glow in the light of an autumn sunset as the pair lead the almost an hour late 1Z37 return 1420 Newquay-Manchester Piccadilly on the approach to Halloon Crossing just to the west of St Columb Road. I should be thankful for fate playing her hand as the picture would not have been so dramatic had the train been on time, late running rail tours can sometimes be a blessing in disguise. For the passengers what better in the declining light and in to the darkness enjoying a ride in Mark 1 stock listening to these fine machines powering their way home? 23 November 1991

So to the year 1992, the only sure way by this time to see the class in action on passenger duties were the rail tours, so I will finish my look at the class 50s from their birth to those final last days by recalling the tours I could manage to get out and see, with a couple of incidentals along the way. The most ambitious of these trips was an exceedingly long day out both in distance (about 750 miles) and endurance (the better part of 20 hours – more stamina back then) from West Devon back to the Northern Fells for the "The Carlisle Fifty Farewell" promoted by Network Southeast. A last chance to see them back where it all began was one not to be missed. So it was back to those green fields and walking the paths of Shap as 50 007 + 50 050 approach the summit passing Shap Wells with that one familiar noise echoing over the hills once again with 1Z38 0820 London Waterloo-Carlisle. Was this the first ever direct train from the SR London terminus to the Border City? It may well have been the first to bring NSE liveried stock to this part of the world far from the lands of the London commuter. 11 April 1992

The general idea of this pilgrimage to Shap for the 50s last fling over the Northern Fells was to try and update a couple of pictures from my exploits there with steam in back in 1967, when of course there were not any electric wires to be seen, and no M6 Motorway either come to that, although I must admit the presence of the latter made it far more realistic to accomplish expeditions like this. On the descent from Greenholme to Tebay where the descent lessens from 1 in 75 to 1 in 146, 50 050 leads 50 007 for the return leg 1655 Carlisle-London Waterloo, the engines nicely fitting on the three arch span of brick and stone Birkbeck Viaduct taking the railway approx. 45 feet over the little stream. So a perfect updated shot of the 1P28 1428 Crewe to Carlisle Parcels with a steam engine at each end taken 22 July 1967 in the bag, all that remained was the 363 mile journey back home, and we almost made it back to Devon in time for a pint. 11 April 1992

By 1992 only 5 members of the class remained in service, mostly on rail tour duty, 50 050 had been returned to its original livery bore its original D400 number, and 50 007 had been repainted into its fifth and final shade of green. The next rail tour I caught up with was actually on the way to see the Steam on the Met, London Transport providing the entertainment on the northern part of the Metropolitan Line. A pairing now more or less standard on such events, 50 050+50 007 both looking in pristine condition on the approach to South Greenford with 1Z10 0800 Aylesbury-Plymouth "Chiltern Atmospheric Tour". Whilst the clientele on board were off to enjoy the delights of my part of the world, I spent a superb day photographing the "Steam on the Met" activities, but that is another story. 16 May 1992

A very notable date and occasion witnessed by many both on the trains and from the lineside, the official last day of class 50 working on the Waterloo-Exeter route with two runs each way Exeter to Salisbury and back, although in reality what would be the short reign of the class 47s on these duties had already been achieved. D400+50 007 head away from Exeter Central passing to the right the former "A" Signal Box with 1035 0928 Exeter St David's-London Waterloo, the view from the Howells Road over bridge. The pair worked the train as far as Salisbury returning with 1V11 1055 London Waterloo-Exeter St David's. Look at all those heads peering from the windows. 24 May 1992

The going away shot from the Howells Road Bridge is included as the two engines gleam in the morning sunshine, the recent repainting of each engine a tribute to the high standards of Laira Shed who put in so much effort to keep the last of the class going. 50 050+50 007 0928 Exeter St David's-London Waterloo, the day bringing to an end the era of the class 50s on the line. There would be a further gathering at this location three weeks later when steam would reappear on the route for the first time since 1966, bear that in mind for the next picture caption. 24 May 1992

Adieu class 50 from the Waterloo line. D400+ 50 007 on the second and final run of the day on a beautiful May afternoon as they pass through the former Seaton Junction with 1041 1543 Exeter St David's-London Waterloo which they worked as far as Salisbury returning from there with 1V17 1655 London Waterloo-Exeter St David's. A sharp contrast to when they passed through earlier with 1V11 when one of the darkest May clouds I have ever encountered ruined the picture. The previously mentioned Shute Arms Hotel can be glimpsed to the far right opposite the station buildings with the farther of the two footbridges packed to capacity, as was the road bridge I was sharing with an equally large throng. Two very nice aged ladies drove past in their little car and stopped to enquire as to what was going on with the usual question" is there a steam train coming?" My reply was that would be in three weeks time, we were all out just practicing for this much anticipated event, and off they went in blissful ignorance. The sun did oblige for the steam three weeks later, with a similar gathering of photographers. 24 May 1992

I did not encounter a lot of class 50 activity in 1993, there was not really much action to be seen in the last full calendar year of the class with just the three celebrities in service, only catching up with them twice on variations of the Atlantic Coast Express. The North Devon weather was at its worst for the Hertfordshire Tours A.C.E. Above we see 50 033 leading and D400 out of sight on the rear passing through Umberleigh in the rain with 1Z41 0835 London Waterloo-Barnstaple. Below: The rain at least had ceased at Lower Braggesford, near Portsmouth Arms for a going away from the camera shot of the 1449 return from Barnstaple, D400 leading and 50 033 on the rear. 18 July 1993

A 14 minute pause at Eggesford for the rail tour to pass the service train on the line gave adequate scope to make my way down to the somewhat isolated Morchard Road Station. The skies were trying to clear as above D400 wends its way through the little station leading 1Z41 1449 Barnstaple-London Waterloo, in the bottom view 50 033 on the rear fits nicely between the telegraph poles with a rather vivid selection of carriage roof colours on show. The last time I looked in here growth had taken control. 18 July 1993

A photographic stop at Copplestone for the tour gave time to move on to Newbridge between Crediton and Cowley Bridge Junction on the long gentle curve there, ideal to portray a rarely seen train of this length on the route at this once fine but now overgrown location. Above the train approaching from Crediton with D400 leading heading towards Exeter with 50 033 on the rear bringing down the curtain on BR operated class 50 activity on The Withered Arm as the former SR system beyond Exeter is coequally known. 18 July 1993

Rare metals as far as class 50s are concerned was the next to come and fortunately the weather to the south of Exeter had much improved to that north of the city. A 20 minute pause at Exeter Central gave enough time for me get ahead of it again, this time to see D400 leading with 50 033 just out of sight on the rear heading away from Polsloe Bridge Halt towards Exmouth as seen from Thorn Park Rise. It had been a long time since a train of this length had been seen on the Exmouth Branch, a view to be treasured due to its rarity and one impossible to replicate today due to the increased timetabled traffic. A half hourly service on the Exmouth Branch almost all day means there is little scope to fit in a rail tour; unless it went down there very early or very late. 18 July 1993

The theory was; after topping and tailing on the Exmouth Branch, and back at Exeter St David's, 50 033 would come off the rear end to run round and double head D400 back up to Waterloo. However, these plans fell in confusion as 50 033 derailed in the running round process at St David's, so rather later than planned we see D400 on its own pulling away from Exeter Central in the late evening sunshine. The normal service trains to Waterloo usually use the up line on the right, so a very rare chance to get the shot from the Howells Road over bridge of a locomotive hauled train using the reversible down line. 18 July 1993

In the process of running round the rail tour and shunting a coach, 50 033 became derailed at Exeter West at the bottom of the incline from Exeter Central, blocking the point allowing access from Platform 1 to reversible down line to Exeter Central. Heads are being scratched, no doubt taxis and buses on order! 18 July 1993

A tour which I helped organise in my professional capacity as a Supervisor in the ticket office at Plymouth station as it was sponsored by local BR management. When the day came I could not get the day off to enjoy a ride or do anything else with it, other than arranging a slightly delayed reporting for duty for the late shift. All very annoying but one has to take these things on the chin, work often got in the way of life- no such problems in retirement. Anyway D400+50 033 look lovely topping the summit of Hemerdon Bank at Sparkwell Bridge with 1Z50 1145 Plymouth-Meldon Quarry "Atlantic Coast Express", and as far as I can recall the first through train from Plymouth to Okehampton since closure of the direct route in May 1968. 25 September 1993

It is getting near the end now, the penultimate rail tour: "The Cornish Caper" was a lengthy day out, travelling west overnight as 1Z50 2115 York-Newquay, the 50s doing the honours west of Bristol. This tour would be the final visit of the class to the Newquay Branch in the BR era, and one where they would break new ground in Cornwall as far as hauling passengers was concerned. After an hour's layover in Newquay where no doubt the early morning cafes did good business, 50 033 leads with 50 050+50 007 out of sight on the rear the 0742 Newquay-Penzance leg of the tour in the trying to break early morning light at Tregoss Summit. Happy memories recalled of Summer Saturdays on the branch. 19 March 1994

Whilst the tour made its squealing way to Par and then reverse I headed west. A lucky burst of light for an angle that could only be done in the early morning as both looking absolutely splendid 50 050+50 007 lead this leg of 1Z50 0742 Newquay-Penzance running beside the A30 dual carriageway at Wheal Busy between Chacewater and Redruth, 50 033 can just be glimpsed to the rear of this lengthy train. Sadly this angle is no longer available to a photographer due to growth of vegetation. 19 April 1994

New ground for the class on passenger duties, and as far as I know only the second time class 50s were seen on the St Ives branch, see 50 018 on page 48. It is full dull weather wise as 50 033 leads the slow approach to Carbis Bay (50 050+50 007 out of sight on the rear) on the 1Z50 1030 Penzance-St Ives leg of this epic tour; so rare the shot had to be done despite the weather, and there was not much chance of doing it again. 19 March 1994

Since the end of steam on the St Ives Branch in September 1961, locomotive hauled trains had been a rarity on the line, and this must count as one of the longest ever, 50 033 out of sight on the front 50 007+50 050 bring up the rear as the cavalcade heads slowly past the station and over Carbis Bay Viaduct. When Boris and his friends gathered here for the G7 meeting in June 2021, a train like this would have brought them here in style. They could not watch the humble sprinter going by as the passenger service on the St Ives branch was suspended for the duration of the conference.
19 March 1994

The Cornish weather was not playing ball with the photography, but who could resist a last fling on the Cornish main line with a pair of large logo liveried 50s. Going well and looking the part 50 033+50 050 have just passed over Tregeagle Viaduct near Probus with the return 1Z50 1310 Penzance-York. 50 007 had to find its own way back to Laira.
19 March 1994

The end of a 26 year era, one I had seen since its beginning in those last days of steam. The 50 Terminator run by Pathfinder Tours was the last train to be class 50 hauled in BR service. Recounting their glories since their migration south from the North West in the early 1970s, this train fittingly ran from London Waterloo to Exeter St Davids, thence to Penzance and back by the direct route to London Paddington.

Work commitments meant I was only going to get one decent picture of this historic occasion, so what better than a classic Cornish Viaduct where they had been part of the scene for 20 years almost to the day. Even the high tide obliged as 50 007 and 50 050 cross Forder Viaduct in the shadow of St Stephen's Church high on the hill behind. This is the Exeter St Davids - Penzance leg of the tour running as 1Z07. 26 March 1994

Remarkably 18 of the class escaped the cutter's torch. It is a sobering thought that these locomotives have been in preservation now for more years than their entire working careers with British Rail. Some can be seen again out on the main line at times, others have graced various heritage railways, and some are in various stages of restoration, overhaul or repair, in need of some tender loving care to erupt in to life again, and of course a variety of liveries some not previously carried during their period of BR ownership; all a good story for another time.

It is another sobering thought that at the time of writing, 30 years have elapsed since the Class 50s finally departed from the British Rail scene, yet through the efforts of the preservationists their fan base continues to grow. This is all due to some remarkable people and organisations such as The Fifty Fund who support the locomotives owned by Class 50 Alliance Ltd, and heritage railways like the South Devon and the Bodmin & Wenford Railways in my part of the world. These groups give enthusiasts and the public the chance of experiencing a Class 50 working a train. For some it will bring back memories and others it will be a chance to savour something new. It certainly never occurred to me when I took that picture of 50 003 at Bodmin Road (now Parkway) on 3 January 1975, that almost half a century later one would still be able to enjoy the sight and sound of a Class 50 coming and going from the station on timetabled trains.

I have been privileged to be able to tell the class 50 story from beginning to end in front line service, and I hope you, the reader, have enjoyed all the memories. What is so reassuring is that through the preservation movement all is not lost. Long live the 50!!

Keeping Class 50s For the Future

The Fifty Fund is the supporters' group for the Class 50 Alliance who own and operate six class 50 locomotives. The Fund's two main objectives are fundraising and education. By purchasing this book you have helped us achieve both objectives. The money raised from the sale of the book will help fund the maintenance of our locomotives and we hope that this book both entertains and gives you some new insight into their operation while owned by British Rail.

The Fund looks to publish a small number of books on the class 50 each year. Each book will cover different aspects, geographical, technical, or operational. They will use a range of authors from different backgrounds to bring a range of perspectives, and hence build a library that tells the full story of these fabulous locomotives. To illustrate these publications we are grateful to the army of photographers who had the foresight to take pictures of the locomotives and are prepared to let us use them in this way.

Below are a number of titles already published and available in bookstores or from our online shop: www.fiftyfundstore.co.uk

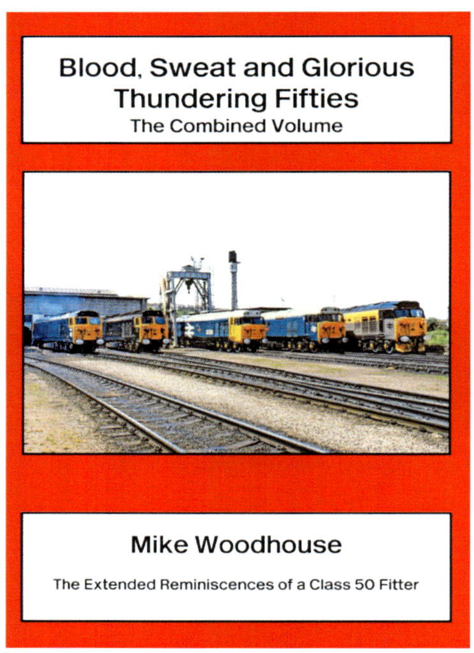

We also publish a calendar each year featuring the best photographs of our locomotives at work around the UK, see the online store for more details.

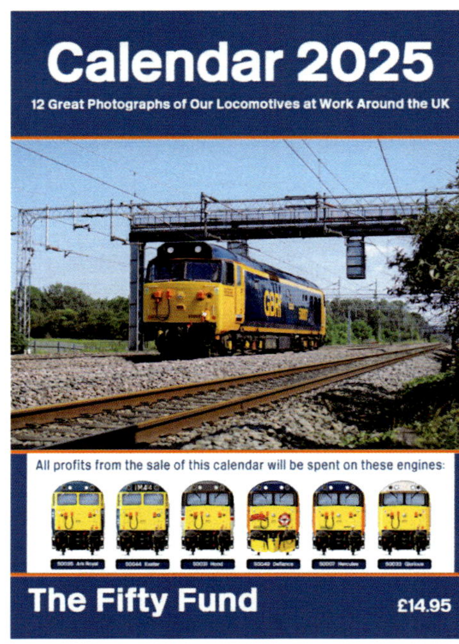

If you'd like to join us, please consider purchasing a share in Class 50 Alliance Ltd: a prospectus is available on our website: www.fiftyfund.org.uk

As well as selling our books, and some similar subject matter titles from other publishers, our on line store also stocks a wide range of railway related merchandise. Including: clothing, replica nameplates, DVDs, models, badges, cards, posters, tea towels, mugs, keyrings, and other miscellaneous railwayana. So it is always worth a visit to see what is new, or to find that perfect birthday present for the rail enthusiast in your life.

Also available from The Fifty Fund

These two books from The Fifty Fund are a very personal view by our founding Chairman.

Dave says about his first book:

You should do a book they said. Finally, after 40 years of encouragement, good intentions, and false starts, my first book has been published.

It tells how a young boy from Plymouth got drawn into Fifty bashing and railway photography, told using railway enthusiast language. A word of warning: This is not a dry historical document about these locos. It light heartedly focusses on "the joined up railway"! that existed after the class was refurbished; it s variety, infrastructure and working practices. The environment in which we went bashing (being hauled by our favourite locos). The Large Logo livery is king in this title, with a nod towards what went before, and after; with the arrival of the bean counters. In reality the (hopefully interesting) storylines are just an excuse to showcase pictures I took over the years. A visual record of "moves" made in pursuit of mileage or rare track with 50s.

There are photos from far flung parts of the country, on tours and unusual workings, including the Track test Train. Along with visual records of days spent trudging round the countryside, standing in fields in all weathers, waiting for something to turn up. The "rules" of railway photography (three quarter views in bright sunshine), may have been studiously ignored and there's some arty stuff too.

Dave says about his follow up book:

You should do a book they said. Then, after 40 years, two come along in quick succession! This latest tome explains how my hobby (bashing and photographing) was stolen. One minute 50s were being refurbished with a long life ahead of them, the next they were being put on the scrap heap. The narrative is told in the same tongue in cheek manner as the first book, using enthusiast's lingo.

As before photos conform to my own railway photography rules. From the outset the class was repetitively used as a stop gap until something else replaced them. But then the search for work ended abruptly. Bean counters dismantled the joined up railway to prepare it for privatisation. Everything was split up and put into cost centres for accountancy purposes. This changed the way the locos were maintained and used. Sectorisation was invented. Using photos and anecdotes the role of the class within each Sector is explained. Without exception this ended in tears. The dubious 50149 project and increased departmental work created new photographic opportunities, but it was sad to see 100mph ETH locos used in this fashion.

Opportunities are taken to muse over what might have been; if 50s had been allocated to InterCity, or if the railway was privatised differently.

The Fifty Fund

www.fiftyfund.org.uk

50035 Ark Royal | 50044 Exeter | 50031 Hood | 50049 Defiance | 50007 Hercules | 50033 Glorious